I0469329

August 2012

BATTERIES AND ENERGY STORAGE

Federal Initiatives Supported Similar Technologies and Goals but Had Key Differences

GAO

Accountability ★ Integrity ★ Reliability

GAO-12-842

G A O
Accountability * Integrity * Reliability

Highlights

Highlights of GAO-12-842, a report to congressional requesters

BATTERIES AND ENERGY STORAGE

Federal Initiatives Supported Similar Technologies and Goals but Had Key Differences

Why GAO Did This Study

Federal interest in batteries and other energy storage technologies has increased in recent years to help address energy, defense, and space exploration challenges. The federal government has devoted substantial resources to support such technologies for the electric grid, electric vehicles, warfighting, and other uses.

GAO was asked to (1) identify the scope and key characteristics of federal battery and energy storage initiatives; (2) determine the extent to which there is potential fragmentation, overlap, or duplication, if any, among these initiatives; and (3) determine the extent to which agencies coordinate these initiatives. GAO focused on fiscal years 2009 through 2012 because DOE made large investments in these technologies during these years. GAO surveyed initiatives identified in six agencies: DOE, DOD, NASA, NSF, EPA, and NIST. GAO included questions about the following key characteristics: funding obligations, technologies, uses, technology advancement activities, goals, eligible funding recipients, and funding mechanisms. GAO analyzed survey responses and interviewed agency officials to gather more information. GAO examined external coordination for all agencies and internal coordination in DOE and DOD because they had the largest obligations of the agencies GAO reviewed.

This report contains no recommendations. In response to the draft report, DOE, DOD, NASA, and NSF provided technical comments, which GAO incorporated as appropriate. The other agencies had no comments.

View GAO-12-842. For more information, contact Frank Rusco at (202) 512-3841 or ruscof@gao.gov.

What GAO Found

GAO identified 39 battery and energy storage initiatives with a variety of key characteristics that were implemented across six agencies: the Departments of Energy (DOE) and Defense (DOD), the National Aeronautics and Space Administration (NASA), the National Science Foundation (NSF), the Environmental Protection Agency (EPA), and the National Institute of Standards and Technology (NIST). These initiatives obligated over $1.3 billion from fiscal years 2009 through 2012. Initiatives supported a variety of technologies, uses, advancement activities, and goals. Several types of recipients were eligible for funding, such as private industry, universities, and federal labs, through contracts, grants, and other mechanisms.

GAO found that initiatives were fragmented and had overlapping characteristics but did not find clear evidence of duplication. Initiatives were fragmented because they were involved in the same broad area of national need: to advance batteries and other energy storage technologies. Thirty initiatives had overlapping characteristics in that they supported similar technologies, uses, advancement activities, and goals. These initiatives also had similar types of eligible funding recipients. Although fragmented and overlapping initiatives create the risk of potential unnecessary duplication, initiatives supported agency-specific missions and strategic priorities that differentiated their efforts. In addition, agency officials involved with the initiatives reported differences in the technologies needed for specific uses, specific goals, and the types of recipients eligible for assistance.

Agencies reported several activities to coordinate with each other on their battery and energy storage initiatives, including initiatives that were overlapping. Activities were consistent with practices that GAO has previously reported can help enhance coordination such as agreeing on roles and responsibilities. In addition, DOE has taken steps to internally coordinate its battery and energy storage initiatives through activities that, among other things, defined common technology goals. DOD has also taken actions to improve its coordination of battery and energy storage initiatives based on a recommendation in a prior GAO report.

Agency Battery and Energy Storage Initiatives and Funding Obligations, Fiscal Years 2009 through 2012

Agency	Number of initiatives	Funding obligations[a]
DOE	11	$851,994,808[b]
DOD	14	430,274,229[c]
NASA	8	20,811,374[c]
NSF	4	8,582,868
EPA	1	3,258,029
NIST	1	1,375,000[c]
Total	**39**	**$1,316,296,308**

Source: GAO analysis of survey responses.

[a] All funding obligations for fiscal year 2012 are estimated.

[b] In addition to these obligations, DOE supported about $596 million in direct loans.

[c] Includes estimates for some obligations for fiscal years 2009 through 2011.

_____ **United States Government Accountability Office**

Contents

Abbreviations

ARPA-E	Advanced Research Projects Agency-Energy
ATVM	Advanced Technology Vehicles Manufacturing
AVPTA	Advanced Vehicle Power Technology Alliance
BEEST	Batteries for Electrical Energy Storage in Transportation
BES	Office of Basic Energy Sciences
CAES	compressed air energy storage
COTS	commercial-off-the-shelf
CRS	Congressional Research Service
CWG	Chemical Working Group
DARPA	Defense Advanced Research Projects Agency
DOD	Department of Defense
DOE	Department of Energy
EERE	Office of Energy Efficiency and Renewable Energy
EPA	Environmental Protection Agency
EPCOI	Energy and Power Community of Interest
ESTCP	Environmental Security Technology Certification Program
GRIDS	Grid-Scale Rampable Intermittent Dispatchable Storage
LPO	Loan Program Office
MOU	memorandum of understanding
NASA	National Aeronautics and Space Administration
NIST	National Institute of Standards and Technology
NSF	National Science Foundation
OE	Office of Electricity Delivery and Energy Reliability
OEPP	Office of the Assistant Secretary of Defense for Operational Energy Plans and Programs
OMB	Office of Management and Budget
OVT	Office of Vehicle Technologies
PASTA	Panel of Senior Technical Advisors
QTR	Quadrennial Technology Review
Recovery Act	2009 American Recovery and Reinvestment Act
TARDEC	Tank Automotive Research, Development and Engineering Center
Treasury	Department of the Treasury
TRL	technology readiness level

August 30, 2012

The Honorable Ralph Hall
Chairman
Committee on Science, Space, and Technology
House of Representatives

The Honorable Andy Harris
Chairman
Subcommittee on Energy and Environment
Committee on Science, Space, and Technology
House of Representatives

Federal interest in batteries and other energy storage technologies has increased in recent years to help address key energy, defense, and space exploration challenges. According to the Department of Energy (DOE), enhancing these technologies contributes to more flexible and efficient control of the nation's existing electric infrastructure, or grid (e.g., electricity networks including power lines and customer meters). For example, these technologies have the potential to facilitate greater use of intermittent renewable energy sources, such as wind and solar energy, on the grid. In addition, according to DOE, batteries are also critical to advancing electric vehicles that are commercially viable to help reduce U.S. oil consumption.[1] To address these challenges, DOE has devoted substantial resources in recent years to initiatives to support batteries and other energy storage technologies. For example, DOE awarded $185 million in funds made available under the 2009 American Recovery and Reinvestment Act (Recovery Act) to 16 projects that supported demonstrations of batteries and other energy storage technologies on the electric grid. In addition, we reported in December 2010 that the Department of Defense (DOD) has provided significant support for batteries and other energy storage technologies in recent years because virtually all DOD weapon systems and equipment rely on these technologies.[2] For example, we reported that DOD invested about $260

---

[1] In this report, the term "electric vehicles," refers to different types of electrified vehicles including hybrid electric, plug-in hybrid electric, and all-electric vehicles.

[2] These technologies are called "power sources" in DOD because some devices' main use is power generation. An example of such a use is a battery supplying power for a laptop computer.

million in science and technology efforts to develop and improve batteries from fiscal years 2006 through 2010.[3] The National Aeronautics and Space Administration (NASA) also relies on and has invested in these technologies to help support space exploration missions because space craft, space stations, and astronaut space suits require power that can be supplied remotely. Several other federal agencies—including the National Science Foundation (NSF), and the Department of Commerce's National Institute of Standards and Technology (NIST)—have also invested in initiatives to support research and development of these technologies.

The existence of these initiatives at multiple agencies has raised questions about the potential for duplication, which in this context occurs when multiple initiatives support the same technology advancement activities for the same technologies and uses, provide funding to the same recipients using the same funding mechanisms, and have the same goals. As we previously reported, unnecessary duplication can potentially result from fragmentation and overlap among government programs.[4] Fragmentation occurs when more than one federal agency (or more than one organization within an agency) is involved in the same broad area of national need. For purposes of this report, overlap occurs when multiple initiatives support similar technologies, uses, technology advancement activities, and funding recipients, and have similar goals. We have previously reported that coordination across programs may help address fragmentation, overlap, and duplication.[5] In this report, we define coordination as any joint activity by two or more organizations that is intended to produce more public value than could be produced when the organizations act alone.

In this context, you asked us to review federal initiatives supporting batteries and other energy storage technologies—which we termed

---

[3]Specifically, we reported that DOD invested about $138 million in science and technology efforts related to nonrechargeable (i.e., primary batteries) and about $122 million in efforts related to rechargeable (i.e., secondary batteries) during fiscal years 2006 through 2010. See GAO, *Defense Acquisitions: Opportunities Exist to Improve DOD's Oversight of Power Source Investments*, GAO-11-113 (Washington D.C.: Dec. 30, 2010).

[4]GAO, *Opportunities to Reduce Potential Duplication in Government Programs, Save Tax Dollars, and Enhance Revenue*, GAO-11-318SP (Washington, D.C.: Mar. 1, 2011).

[5]GAO-11-318SP and GAO, *Employment For People With Disabilities: Little Is Known about the Effectiveness of Fragmented and Overlapping Programs*, GAO-12-677 (Washington, D.C.: June 29, 2012).

GAO-12-842 Batteries and Energy Storage

"federal battery and energy storage initiatives." Our objectives were to (1) identify the scope and key characteristics of federal battery and energy storage initiatives; (2) determine the extent to which there is potential fragmentation, overlap, or duplication of these initiatives, if any; and (3) determine the extent to which agencies coordinate their battery and energy storage initiatives.

To address these three objectives, we focused our review on rechargeable batteries[6] and certain other energy storage technologies; we excluded nonrechargeable batteries,[7] fuel cells, and nuclear energy storage technologies. We also focused on initiatives active[8] during fiscal years 2009 through 2012 because, during these years, DOE offices made large investments in these technologies, including with funds made available under the Recovery Act. For example, DOE's Advanced Research Projects Agency-Energy (ARPA-E) supported about $97 million in battery and energy storage technologies during fiscal years 2010 and 2011 with funding made available under the Recovery Act. We defined an initiative as a group of agency activities serving a similar purpose or function, such as a program or mission area. We included initiatives that either exclusively supported battery and energy storage technology projects or did so as part of a broader effort that also supported other types of projects.[9] For example, we included DOE's Title XVII Loan Guarantee Program, which supported loans to commercial-scale renewable energy projects, including battery, solar generation, wind generation, and geothermal generation projects.

- To identify the scope and key characteristics of federal battery and energy storage initiatives, we first identified potentially relevant agencies and initiatives by reviewing two previous GAO reports that collected information on some federal battery and energy storage

---

[6]Rechargeable batteries can be reenergized after their charge has been depleted.

[7]Nonrechargeable batteries are discarded after their charge has been depleted.

[8]For the purposes of this report, we defined active initiatives as those that were planned or funded or implemented or authorized in any of the fiscal years described.

[9]For the purposes of this report, we defined individual projects as parts of an overall initiative—for example, specific grant awards, agreements, in-house research activities, or contracts supported by an initiative.

initiatives active in fiscal year 2010.[10] In addition, we searched key federal databases, and interviewed agency officials. We limited our scope to initiatives that supported basic science, applied research, demonstrations, and commercialization activities for batteries and other energy storage technologies. Because initiatives often supported more than one type of technology advancement activity, some of the initiatives included in this report may also support deployment activities—efforts to facilitate or achieve widespread use of technologies either in the commercial market or for federal agencies' use. However, in identifying agencies and initiatives, we excluded initiatives that focused solely on deployment activities and initiatives involving agency-owned assets such as fleets or facilities. For example, we did not include the Department of the Treasury (Treasury) and several federal tax credits it administered that indirectly supported deployment of batteries and other energy storage technologies during fiscal years 2009 through 2012. Treasury officials told us that data were generally not available on the estimated revenue loss directly associated with tax credits for these technologies because available data do not break out qualifying investments in fine enough detail to determine which part may have been for qualifying battery and energy storage devices. We confirmed a final list of six agencies and 39 initiatives. We then conducted a survey of agency officials involved in each of the 39 initiatives and included questions about the following key characteristics: funding obligations; goals and performance measures for batteries and other energy storage technologies; types of technologies, uses, and advancement activities supported; types of eligible funding recipients and funding mechanisms used to provide assistance; and coordination activities. Some officials reported that funding obligations for fiscal years 2009 through 2011 were estimated because battery and energy storage projects were portions of larger projects, or broader initiatives, and were not specifically tracked. In addition, all funding obligations reported for fiscal year 2012 were estimated. We received survey responses from all 39 initiatives and therefore had a response rate of 100 percent. According to an official from DOE's Office of Energy Efficiency and Renewable Energy's (EERE) Office of Vehicle Technologies (OVT), since the survey focused on research and development activities, they did not include in their survey

---

[10]GAO-11-113 and GAO, *Renewable Energy: Federal Agencies Implement Hundreds of Initiatives*, GAO-12-260 (Washington, D.C.: Feb. 27, 2012).

response $1.5 billion in funds made available under the Recovery Act. With these funds DOE made awards to 20 projects to support the establishment of advanced battery manufacturing and battery recycling facilities in the U.S. DOE officials provided this information separately in OVT's annual energy storage research progress reports.[11]

- To determine the extent of potential fragmentation, overlap, or duplication, if any, we analyzed survey responses for the initiatives to identify similarities and differences in their key characteristics. We then interviewed agency officials to gather more information to further evaluate similarities and differences.

- To determine the extent to which agencies coordinated their initiatives, we used survey responses and interviews to identify interagency coordination activities across the six agencies. We also examined internal coordination in DOE and DOD of initiatives within those agencies because they had the largest funding obligations among the agencies we reviewed. For DOD we followed up on actions, if any, DOD took to address recommendations we made in 2010. We drew on past GAO work related to interagency coordination.[12]

We provide a more in-depth discussion of our questionnaire and methods in appendix I. For a copy of our questionnaire, see appendix V.

We conducted this performance audit from September 2011 to August 2012 in accordance with generally accepted government auditing standards. Those standards require that we plan and perform the audit to obtain sufficient, appropriate evidence to provide a reasonable basis for our findings and conclusions based on our audit objectives. We believe that the evidence obtained provides a reasonable basis for our findings and conclusions based on our audit objectives.

---

[11]See, for example, Department of Energy, Energy Efficiency and Renewable Energy, *Fiscal Year 2011 Annual Progress Report for Energy Storage R&D*, DOE/EE-0675 (Washington, D.C.: Jan. 2012).

[12]GAO, *Results-Oriented Government: Practices That Can Help Enhance and Sustain Collaboration among Federal Agencies*, GAO-06-15 (Washington, D.C.: Oct. 21, 2005).

Background

Batteries and other energy storage technologies include numerous devices that can store energy in one form—such as chemical, mechanical, electrostatic, or thermal energy—and can transform the energy to generate electrical power at a later time. For example, batteries store energy in chemical form and convert it to electrical energy through electrochemical processes. Other types of energy storage technologies include capacitors and flywheels. Capacitors store energy in electrostatic form and release the energy as electrical power. Flywheels store energy in mechanical form using a spinning wheel, or tube, and then transform the energy to electrical power using the spinning wheel, or tube, to drive a generator. Appendix II provides descriptions of technologies identified during our review.

Batteries and other energy storage technologies can support diverse uses, such as the following:

- *Personal-use electronics power.* Internal power for small or mobile electronic devices such as laptop computers and cell phones.

- *Stationary power storage.* Energy storage for stationary electricity generation and distribution systems such as the national electric grid, or stand-alone systems for military installations.

- *Vehicle propulsion.* Power for propulsion of vehicles that travel on the ground, in the air, in and on water, or in space.

- *Auxiliary power for vehicles.* Power for uses other than propulsion such as providing electricity for navigation, communication, and other equipment.

- *Weapon systems power.* Power for weapons and their components necessary for their operation, such as targeting and guidance devices.

Batteries and other energy storage technologies are in various stages of technology advancement. For example, lithium-ion batteries are a popular type of battery used for powering most small or mobile consumer electronics. Recently, lithium-ion batteries are also being used in electric grids and electric vehicles because, among other factors, they are lightweight and have high energy and power densities compared with other commercially available batteries.[13] However, research efforts on lithium-ion batteries for these other uses aim to, among other things, improve their safety and reduce their cost. For example, DOE supports research on lithium-ion batteries used in electric vehicles that focuses on developing new approaches for managing the temperature ranges in which the batteries can operate to help improve their safety.

Federal agencies support innovation of battery and energy storage technologies through a spectrum of technology advancement activities. For purposes of this report, we call these activities basic research, applied research, demonstrations, commercialization, and deployment activities and define these terms as follows:[14]

- *Basic research* includes efforts to explore and define scientific or engineering concepts, or is conducted to investigate the nature of a subject without targeting any specific technology.

- *Applied research* includes efforts to develop new scientific or engineering knowledge to create new and improved technologies.

- *Demonstrations* include efforts to operate new or improved technologies to collect information on their performance and assess readiness for widespread use.

---

[13]Energy density of a battery or other energy storage technology is the ratio of energy stored to the mass or volume of the device. For example, energy density of a battery affects the distance a vehicle can travel with a given size battery. Power density is the amount of energy in a given device's mass or volume that can be delivered in a given period of time. For example, power density of a battery affects how fast a vehicle can accelerate.

[14]We developed definitions that could be applied broadly to make comparisons across agencies and that covered the full spectrum of advancement activities. Federal agencies use various definitions and categories for describing the stages of technology advancement. For example, NASA and DOE use technology readiness level (TRL) categories and definitions to measure and communicate technology readiness for first-of-a-kind technology applications. However, these agencies' TRL categories and definitions are not the same. In addition, the Office of Management and Budget (OMB) provides federal definitions for the terms basic research, applied research, and development in OMB Circular No. A–11, Section 84—Character Classification (Schedule C), but does not provide definitions for demonstrations and commercialization activities. During pretests of our questionnaire, agency officials were able to fit their initiatives' activities within our categories and based on our definitions, which validated our use of them.

- *Commercialization* includes efforts to bridge the gap between research and demonstration activities, and venture capital funding and marketing activities, through transitioning technologies to commercial applications.

- *Deployment activities* include efforts to facilitate or achieve widespread use of technologies either in the commercial market or for federal agencies' use.

Federal agencies pursue technology advancement activities through the direct work of agency staff (i.e., in-house) or by providing funding assistance to external recipients in government laboratories, universities, industry, and nonprofit organizations. Government laboratories include facilities within a federal agency such as the Army Research Laboratory and Air Force Research Laboratory. Such laboratories also include federally funded research and development centers sponsored by federal agencies, such as DOE's national laboratories, that are not federal entities but that were originally established by federal authority to address national scientific research needs.

Six Agencies Supported Thirty-Nine Battery and Energy Storage Initiatives with a Variety of Key Characteristics

We identified 39 battery and energy storage initiatives across six agencies that supported battery and energy storage technologies. In total, these initiatives obligated over $1.3 billion during fiscal years 2009 through 2012 to support a variety of goals, technologies, uses, and advancement activities. They also provided funding to many types of recipients, such as federal labs and private industry, through a number of different funding mechanisms. We provide descriptions of each of these initiatives in appendix III and selected survey responses for each initiative in appendix IV.

Agencies Involved, Initiatives, Funds Obligated, and Goals

We identified 39 initiatives—implemented by six agencies—that supported batteries and other energy storage technologies to some degree.[15] As shown in table 1, over 80 percent of these initiatives (33 of

---

[15]For the purposes of this report, we defined an initiative as a group of agency activities serving a similar purpose or function, such as a program or mission area. Given this broad definition, we included initiatives that did not exclusively support battery and energy storage technology projects but did so as part of a broader effort that also supported other types of projects.

39) were implemented by three agencies: DOE (11), DOD (14), and NASA (8). Other initiatives were implemented by NSF, EPA, and NIST. Twenty-six initiatives were active during all four fiscal years we examined—2009 through 2012—and three initiatives were new in fiscal year 2012.

Overall, as shown in table 1, the six agencies reported that they obligated over $1.3 billion during fiscal years 2009 through 2012 to support these initiatives, according to our analysis of survey responses. Obligations for these initiatives rose from about $246 million in fiscal year 2009 to about $311 million in fiscal year 2012. DOE and DOD had the largest obligations among the six agencies.

- DOE obligated almost $852 million over the period for its 11 initiatives. The largest DOE initiatives were the OVT's Vehicle Technologies Energy Storage Research and Development initiative, which reported obligating $323 million from fiscal year 2009 through 2012;[16] and the Office of Electricity Delivery and Energy Reliability's (OE) Energy Storage Program, which reported obligating $241 million from fiscal year 2009 through 2012, including $185 million in obligations from funding made available under the Recovery Act. DOE's total obligations also included about $30 million in credit subsidy costs obligated by the Loan Program Office's (LPO) Advanced Technology Vehicles Manufacturing (ATVM) Loan Program to support about $520 million in loans that helped establish two electric vehicle battery manufacturing facilities, as well as about $11 million in credit subsidy costs obligated by the Title XVII Loan Guarantee Program to guarantee about $76 million in loans for battery and energy storage

---

[16]DOE's OVT also awarded $1.5 billion in funds made available under the Recovery Act to 20 projects to support the establishment of advanced battery manufacturing and recycling facilities in the United States.

technology projects.[17] Credit subsidy costs are the government's estimated net long-term cost, in present value terms, of the loans.[18]

- DOD reported obligating about $430 million for its 14 initiatives. The DOD initiative with the largest obligations was the Navy and Marine Corps Energy Storage System Integration initiative, which obligated $130 million from fiscal year 2009 through 2012. The next largest DOD initiatives were the Navy and Marine Corps Energy Storage Science and Technology initiative and the Army Energy Storage for Air and Ground Vehicles initiative, which both obligated about $80 million between fiscal year 2009 and 2012.

Table 1: Agency Battery and Energy Storage Initiatives and Funding Obligations, Fiscal Years 2009 through 2012

Agency	Number of initiatives	Funding obligations
DOE	11	$851,994,808[a]
DOD	14	430,274,229
NASA	8	20,811,374
NSF	4	8,582,868
EPA	1	3,258,029
NIST	1	1,375,000
Total	**39**	**$1,316,296,308**

Source: GAO analysis of survey responses.

Note: Some of these totals included estimated obligations for fiscal years 2009 through 2011. All obligations for fiscal year 2012 were estimated by agencies.

[a]In addition to these obligations, DOE supported about $596 million in federal loans for battery and energy storage technology projects.

---

[17]Officials from DOE's LPO told us that one of the battery and energy storage projects with a loan guarantee commitment for a loan of $17 million withdrew from the program reducing the value of its loan to zero. As a result, the total amount in loans that DOE guaranteed for these technologies under the Title XVII Loan Guarantee Program was reduced by $17 million and the total amount of the program's credit subsidy obligations were also reduced by over $830,000.

[18]Credit subsidy costs exclude administrative costs and any incidental effects on governmental receipts or outlays. Present value is the worth of the future stream of returns or costs in terms of money paid immediately. In calculating present value, prevailing interest rates provide the basis for converting future amounts into their "money now" equivalents.

GAO-12-842 Batteries and Energy Storage

Agencies reported that 26 of the initiatives had goals directly related to advancing batteries and other energy storage technologies. For example, DOD's Army Energy Storage for the Soldier initiative aimed to achieve specific targets for increasing the energy density and reducing the weight of devices that soldiers carry. The other initiatives that supported batteries and other energy storage technologies did not have goals directly related to advancement of these technologies, according to their responses. For example, seven initiatives—from DOE, DOD, and NSF—told us that they were focused on basic science research. Five other initiatives reported that they supported batteries and other energy storage technologies as part of a broader mission and, therefore, their goals were not specifically related to advancement of these technologies. For example, DOE's ATVM Loan Program reported that its mission was to provide loans to companies making cars and components in U.S. factories that increase fuel economy at least 25 percent above the fuel economy levels for 2005. Accordingly, the program provided loans to two companies to manufacture battery packs for electric cars. This program also provided loans for a variety of other projects, such as upgrading factories and improving vehicle fuel efficiency for vehicles powered by internal combustion engines.[19]

The 26 initiatives that agency officials reported tracked goals directly related to advancing batteries and other energy storage technologies reported using various technical performance and cost targets. For example, DOE's OE Energy Storage Program established a number of technical performance targets related to improving the cost, cycle life, and energy efficiency of batteries; flywheels; compressed air energy storage; and other technologies for providing stationary storage on the electric grid. NASA's Flywheel Energy Storage and Momentum Control initiative also established technical performance targets to develop a flywheel that can discharge the same amount of energy for 3 hours as the best field-deployed battery systems. In addition, DOD's Army Energy Storage for the Soldier initiative set a cost target to produce a battery that matches the contour of a soldier's body armor and that costs less than $400.

---

[19]In addition, DOE's Batteries and Energy Storage Energy Innovation Hub reported not having battery and energy storage related goals. This initiative was established in fiscal year 2012. At the time of our review, DOE officials from the Office of Science's Office of Basic Energy Sciences (BES) reported that the applications and award process was not completed. The goals for the hub will be proposed through the applications, reviewed by peer review, and finalized in negotiations between DOE and the successful applicant.

Technologies, Uses, and Technology Advancement Activities

As shown in table 2, agencies reported that their initiatives supported more than 10 technologies, with most initiatives supporting two or more technologies. Specifically, 21 of the 39 initiatives supported more than one kind of battery or other energy storage technology, and initiatives supported on average two technologies. Lithium-ion batteries were the most commonly supported technology among initiatives. Many initiatives also reported that they supported other types of batteries, such as metal-air batteries, lithium-metal batteries, and other energy storage technologies, such as capacitors, flywheels, and compressed air energy storage (CAES). Appendix II provides descriptions of batteries and other energy storage technologies identified in this review.

Table 2: Number of Initiatives Supporting Each Type of Technology

Technology	Number of initiatives
Lithium-ion batteries	28
Metal-air batteries	19
Capacitors	17
Lithium-metal batteries	16
Basic energy storage research	14
Advanced lead-acid batteries	11
Redox flow batteries	9
Sodium batteries	9
CAES	4
Flywheels	4
Other[a]	15

Source: GAO analysis of survey responses.

Note: Numbers total more than 39 because many initiatives supported more than one type of technology.

[a]According to questionnaire responses, other technologies supported by initiatives included, for example, hydraulic accumulators, superconducting magnetic energy storage, solar thermal energy storage, and other types of flow batteries such as zinc-bromide flow batteries.

Agencies reported that 32 of the 39 initiatives supported specific technology uses. Of these 32 initiatives, 24 reported that they supported more than one technology use category; on average, initiatives supported two technology use categories. As shown in table 3, the largest number of initiatives focused on ground-based vehicle propulsion, such as for light-duty electric vehicles and space rovers. The next largest groups of initiatives focused on auxiliary power for vehicles, which is power for uses such as starting and stopping vehicles or supporting electronics in military tactical vehicles, as well as on stationary power storage, including large-scale systems such as the electric utility grid and small-scale systems such as microgrids providing power for forward operating bases.[20]

Table 3: Number of Initiatives Supporting Battery and Energy Storage Technology Uses

Technology uses	Number of initiatives
Ground-based vehicle propulsion	16
Auxiliary power for vehicles	14
Stationary power storage	14
Propulsion for vehicles that are not ground-based (air, space, underwater)	12
Personal-use electronics power	6
Weapon systems	6
Other[a]	8

Source: GAO analysis of questionnaire responses.

Note: Numbers total more than 39 because many initiatives supported more than one type of technology use.

[a]Other types of technology use—as marked in questionnaire responses—include, for example, space applications and low-power sensors for defense applications.

Agencies' initiatives supported one or more of the five categories of technology advancement activities. Specifically, for 28 of the 39 initiatives, agency officials reported supporting more than one technology advancement activity. As shown in table 4, the most common type supported was applied research, followed by demonstrations, basic research, commercialization, and deployment activities. Agencies reported that initiatives that supported applied research focused on a broad range of research that included developing practical devices—such as a battery for an electric vehicle—to meet specific needs, designing energy and power system integration and management approaches, and developing new or improved manufacturing processes. Agencies reported that initiatives that supported basic research generally targeted discovery and development of new energy storage materials and approaches for device architectures.

---

[20]A forward operating base is an airfield or ground base used as a staging area to support tactical operations without establishing full support facilities.

Table 4: Number of Initiatives Supporting Each Technology Advancement Activity

Technology advancement activity	Number of initiatives
Basic research	15
Applied research	27
Demonstrations	19
Commercialization	12
Deployment[a]	10
Other[b]	6

Source: GAO analysis of questionnaire responses.

Note: Numbers total more than 39 because many initiatives supported more than one type of advancement activity.

[a]Initiatives that were surveyed did not solely support deployment but rather did so along with other technology advancement activities. We report these deployment activities that initiatives identified as part of their initiatives but these do not encompass the entirety of federal deployment activities.

[b]Other types of technology advancement activities—as marked in questionnaire responses—include, for example, developing testing procedures.

Types of Recipients and Funding Mechanisms

For 33 of the 39 initiatives, agency officials reported providing funding assistance to external recipients. Most initiatives reported having more than one eligible type of external funding recipient. In addition, 18 initiatives reported conducting in-house technology advancement activities. As shown in table 5, officials commonly reported that industry, universities, and DOD laboratories were eligible types of funding recipients, along with DOE national laboratories, and other federal laboratories. The DOE national laboratories are an important recipient for DOE programs because several of the national laboratories host multidisciplinary battery and energy research centers. For example, according to DOE officials, the national laboratories, among other things, implement one of BES's Energy Frontier Research Centers focused on basic energy storage research. The national laboratories also conduct much of the early applied research supported by OVT's Vehicle Technologies Energy Storage Research and Development initiative and OE's Energy Storage Program.

Table 5: Number of Initiatives Supporting Each Type of Eligible Recipient

Recipient type	Number of initiatives
Industry	32
Universities	30
Department of Defense laboratories	25
Department of Energy national laboratories	23
Other federal government laboratories[a]	23
Other[b]	8

Source: GAO analysis of questionnaire responses.

Note: Numbers total more than 39 because many initiatives had more than one type of eligible recipient.

[a]Agencies reported other federal government labs to include research agencies such as NIST and NSF.

[b]Other types of recipients—as marked in questionnaire responses—include, for example, nonprofit organizations.

For most agencies' initiatives, officials reported using more than one type of funding mechanism to provide assistance. In particular, officials reported funding assistance was provided primarily through three types of mechanisms—contracts,[21] grants, and cooperative agreements.[22] As shown in table 6, officials for 21 initiatives reported using contracts, officials for 18 reported using grants, and officials for 15 reported using cooperative agreements.

---

[21]"Contract" means a mutually binding legal relationship obligating the seller to furnish the supplies or services (including construction) and the buyer to pay for them. It includes all types of commitments that obligate the government to an expenditure of appropriated funds and that, except as otherwise authorized, are in writing. In addition to bilateral instruments, contracts include (but are not limited to) awards and notices of awards; job orders or task letters issued under basic ordering agreements; letter contracts; orders, such as purchase orders, under which the contract becomes effective by written acceptance or performance; and bilateral contract modifications. Contracts do not include grants and cooperative agreements covered by 31 U.S.C. 6301, et seq.

[22]Like grants, cooperative agreements involve the provision of financial or other support to accomplish a public purpose of support or stimulation authorized by federal statute. However, cooperative agreements differ from grants in terms of agency involvement, supervision, and intervention in the project. Whereas grants restrict government involvement to the minimum necessary to achieve program objectives, under cooperative agreements, the government and prime recipients share responsibility for the management, control, direction, and performance of projects.

Table 6: Number of Initiatives Using Each Funding Mechanism

Funding mechanism	Number of initiatives
Contracts	21
Grants	18
Cooperative agreements	15
Interagency agreements	7
Direct loans	2
Other[a]	8

Source: GAO analysis of questionnaire responses.

Note: Numbers total more than 39 because many initiatives used more than one type of funding mechanism.

[a]Other types of funding mechanisms—as marked in questionnaire responses— include, for example, Space Act agreements. These are legal agreements other than contracts, leases, and cooperative agreements that NASA uses under authority granted to it in the National Aeronautics and Space Act of 1958 to give the agency greater flexibility in achieving its mission.

Agencies' Initiatives Were Fragmented and Had Overlapping Characteristics but Did Not Demonstrate Clear Instances of Duplication

According to our analysis, initiatives were fragmented across multiple agencies and had overlapping characteristics, but we found no clear instances of duplicative initiatives. Initiatives were fragmented because they were implemented across six agencies and were involved in the same broad area of national need: to advance new and improved batteries and other energy storage technologies. We also found 30 initiatives in four of the agencies had overlapping characteristics, to some degree, in that they supported broadly similar technologies, uses, technology advancement activities, and goals. These initiatives also generally reported that similar types of recipients were eligible to receive assistance. Fragmented and overlapping initiatives across these agencies resulted from federal efforts to both create and expand programs to improve these technologies for a range of agency missions. Although the existence of fragmented and overlapping initiatives creates the risk of potential unnecessary duplication, initiatives we reviewed supported agency-specific missions and strategic priorities that differentiated them. In addition, initiatives reported differences in the technologies needed for specific uses, specific goals, and the types of recipients they provide support to, and we did not find clear evidence of duplicative initiatives.

Initiatives Were Fragmented and Had Overlapping Characteristics

The 39 initiatives we identified were fragmented because they were implemented by various offices in six agencies and were involved in the same broad area of national need: to advance batteries and energy storage technologies. In addition, we identified 30 initiatives in four of the six agencies—DOE, DOD, NASA, and NSF—that had overlapping characteristics, to some degree, with at least one other initiative in that they supported broadly similar technologies,[23] uses, technology advancement activities, and goals. In addition, these initiatives generally reported similar types of eligible funding recipients. The following are several examples of such overlapping characteristics:

- *Basic science research on energy storage.* Two initiatives supported similar goals for basic research on energy storage. Specifically, in fiscal year 2012, DOE's BES established a new initiative called the Batteries and Energy Storage Energy Innovation Hub that had a goal to rapidly drive toward electrochemical energy storage solutions beyond the current limits, for uses in electric cars and the electric grid. According to DOE officials, the planned hub will consist of a single large multi-year, multi-institution, and cross-disciplinary award with each participant expected to work in a coordinated effort. NSF also established a new initiative in fiscal year 2012, called the Energy for Sustainability Program that the agency reported had a goal to, among other things, support basic research on radically new battery systems or breakthroughs for use in electric cars. An official from NSF told us that awards for the Energy for Sustainability Program will be single-investigator grants made generally to academic researchers.

- *Applied research on ground-based vehicle propulsion.* Three initiatives reported similar goals for applied research on similar batteries for ground-based vehicle propulsion. Specifically, DOE's OVT Vehicle Technologies Energy Storage Research and Development initiative supported applied research to develop next generation battery and capacitor materials, cells, packs, and manufacturing processes to enable a large market penetration of electric vehicles. DOE's ARPA-E Batteries for Electrical Energy Storage in Transportation (BEEST) initiative aimed to develop transformational advanced battery chemistries, architectures, and manufacturing processes that, if successful, would leapfrog next generation technologies and speed widespread adoption of electric

---

[23]In this context, the term "technologies" includes basic energy storage science.

vehicles. DOD's Army Energy Storage for Air and Ground Vehicles initiative focused on applied research to develop battery systems and battery integration approaches that provide hybrid electric vehicle propulsion for Army tactical and other vehicles.

- *Applied research for stationary storage systems.* Three initiatives reported similar goals for applied research to improve batteries and other energy storage technologies for large and small-scale stationary power storage. Specifically, DOE's OE Energy Storage Program focused on developing batteries and other energy storage technologies and systems that will increase the reliability, performance, and competitiveness of electricity generation and transmission in the electric grid and in stand-alone grids. DOE's ARPA-E Grid-Scale Rampable Intermittent Dispatchable Storage (GRIDS) initiative aimed to develop new and transformational cost competitive batteries and other energy storage technologies to increase electric grid reliability and enable greater integration of energy generation from renewable sources, such as wind and solar energy, on the electric grid. DOD's Navy and Marine Corps Energy Storage Science and Technology initiative focused on developing battery and other energy storage devices and systems to enable the Navy and Marine Corps to successfully carry out current and future missions and to retain superiority over adversaries. The initiative's research on these technologies for stationary energy storage focused on developing solutions for providing back-up and hotel power[24] for naval ships, and for enabling the use of alternative energy sources, such as solar or wind energy, to provide continuous and reliable power for Marine forward operating bases and other Navy and Marine facilities.

- *Demonstrations for stationary storage systems.* Two initiatives reported similar goals for demonstrations of batteries and other energy storage technologies to improve the efficiency of stationary power systems and their ability to use renewable energy sources. Specifically, DOD's Installation Energy Test Bed initiative in the Office of the Deputy Under Secretary of Defense for Installations and Environment's Environmental Security Technology Certification

---

[24]Hotel power refers to power provided for all loads on a naval ship other than that to propel the ship through the water (propulsion). Examples of hotel loads are weapon systems, radar systems, and loads needed for the sustainment and comfort of the people on board.

Program had a mission to demonstrate battery technologies for small-scale stationary power uses, such as in fixed DOD installations, that will enable compatibility with the electric grid, improve efficiency of an installation's power network, and enable increased use of distributed energy generation, especially renewable energy sources.[25] DOE's OE Energy Storage Program supports activities to demonstrate the value of and evaluate the performance of batteries and other energy storage technologies and systems on the electric grid, including large-scale and distributed systems, such as community energy storage, to help define electric grid requirements for batteries and other energy storage technologies, and provide valuable knowledge for further development and optimization of these technologies.

In addition, 29 of the 30 broadly overlapping initiatives in DOE, DOD, NASA, and NSF also reported that one or more similar types of recipients were eligible to receive funding assistance. For example, officials from DOE's OVT Vehicle Technologies Energy Storage Research and Development initiative and DOD's Army Energy Storage for Air and Ground Vehicles initiative reported similar types of eligible recipients: industry, universities, DOD laboratories, DOE laboratories, and other federal laboratories.

Fragmented and overlapping initiatives have resulted from federal efforts to both create and expand programs across agencies to improve batteries and other energy storage technologies for a range of agency missions. Officials from DOE and DOD told us that their initiatives had broadly similar characteristics because energy storage technologies are generally enabling technologies that help agencies to work toward broader agency technology goals. For example, DOE is focused on developing batteries to incorporate in electric cars and for use in the electric grid while DOD is focused on incorporating these technologies in military vehicles, soldier worn or carried equipment, and weapons. Officials from DOE, DOD, and NASA also told us that initiatives that have broadly similar characteristics can result in complementary research. For example, the DOE official who leads the OVT's Hybrid and Electric Systems Team told us that DOE has

---

[25]Distributed energy generation refers to a variety of small, modular power-generating technologies such as wind turbines and solar panels that can be combined with energy storage technologies to improve the quality and/or reliability of the electricity supply. They are "distributed" because they are placed at or near the point of energy consumption, unlike traditional "centralized" systems, where electricity is generated at a remotely located, large-scale power plant and then transmitted down power lines to the consumer.

funded the development of a battery abuse modeling tool that predicts how a battery will respond when subjected to abnormally high operating temperatures. DOE officials told us that Army's Tank Automotive Research, Development and Engineering Center (TARDEC) has funded research to adapt the modeling tool to analyze the performance of batteries used in Army systems and military operational scenarios. In addition, officials from DOD told us that overlap can be beneficial in the area of basic research where there are potentially multiple approaches for achieving similar goals. For example, one DOD official said that there are many approaches to developing new energy storage materials and increasing the surface area of these materials at the nanoscale; and having multiple efforts focused on this and similar scientific challenges increases the chances of making these new discoveries. However, the existence of broadly overlapping initiatives also creates the risk of potential duplication.

Federal Initiatives Did Not Demonstrate Clear Instances of Duplication

Fragmentation and overlap are, by themselves, not an indication that unnecessary duplication of initiatives exists. In particular, we found that agencies' battery and energy storage initiatives supported agency-specific missions and strategic priorities that differentiated them. The following are agency missions and strategic priorities supported by each federal agency included in our review:

- DOE's mission is to ensure America's security and prosperity by addressing its energy, environmental, and nuclear challenges through transformative science and technology solutions. DOE's battery and energy storage initiatives support its mission by supporting the agency's strategic plan goal to catalyze the timely, material, and efficient transformation of the nation's energy system and secure U.S. leadership in clean energy technologies such as electric vehicle batteries.
- DOD has defined energy security as a strategic priority for the military services. Energy security involves helping to assure access to reliable supplies of energy and the ability to protect and deliver sufficient energy to meet operational needs. DOD's battery and energy storage initiatives help support this strategic priority by delivering power for equipment carried or worn by soldiers, weapon systems, and military vehicles, as well as enabling tactical operations and fixed military bases to more efficiently use fossil fuels and make greater use of renewable energy sources.
- NASA's mission is to drive advances in science, technology, and exploration to enhance knowledge, education, innovation, economic

vitality, and stewardship of Earth. NASA's battery and energy storage initiatives support its mission by contributing to innovation of space exploration technologies to help achieve its space exploration missions.

- NSF's mission is to promote the progress of science; to advance the national health, prosperity, and welfare; to secure the national defense; and for other purposes. To implement its mission, NSF supports research and education across all fields and disciplines of science and engineering and at all levels of education. NSF-funded fundamental research and education projects fuel innovation and contribute directly to addressing national challenges, such as the development of a clean energy economy. As a result, NSF has supported research on the conversion, storage, and distribution of diverse power sources as well as on energy materials, energy use, and energy efficiency.

- EPA's mission is to protect human health and the environment. EPA's energy storage initiative helps support its mission by supporting improved fuel economy and reduced vehicle air pollution emissions.

- NIST's mission is to promote U.S. innovation and industrial competitiveness by advancing measurement science, standards, and technology in ways that enhance economic security and improve our quality of life. NIST has a single project related to battery and energy storage, which supports this mission by developing new tools and measurement systems needed to characterize the nanoscale properties of materials and devices relevant to energy storage.

We also found that agency officials involved in the initiatives reported differences in the technologies needed for specific uses, their initiatives' specific goals, and the types of recipients eligible for assistance; further differentiating their respective initiatives. Examples are as follows:

- *Differences in technologies for specific uses.* Agency officials involved in broadly overlapping initiatives reported that, although the initiatives may support similar technologies, there were meaningful differences in them because specific uses have unique operational requirements. For example, both DOE's OE Energy Storage Program and DOD's Navy and Marine Corps Energy Storage Science and Technology initiative supported applied research on similar types of batteries for stationary power storage. However, DOE's initiative focused on addressing operational requirements of batteries for use in the commercial electric grid, and DOD's Navy and Marine Corps initiative focused on the operational requirements unique to Navy and Marine Corps-specific uses such as back-up and hotel power on naval ships. For example, officials from DOE's OE told us the operational requirements for using batteries on the commercial electric grid include, among other factors, low cost and long cycle life, which is the amount of time that a battery can operate at a specified level of performance. In contrast, DOD officials from the Navy and the Office of the Assistant Secretary of Defense for Research and Engineering told us that batteries for providing back-up and auxiliary power for naval ships at sea must be light and compact enough to be easily transported, be rugged enough to fully operate after absorbing battlefield shocks, and be capable of operating under extreme temperature ranges such those present inside a naval ship.

- *Differences in specific goals.* Agency officials involved in broadly overlapping initiatives generally reported differences in their initiatives' specific goals. For example, both DOE's OVT Vehicle Technologies Energy Storage Research and Development initiative and DOE's ARPA-E BEEST focused on improving battery technologies to help expand markets for electric vehicles. However, officials from OVT reported that their initiative primarily focused on developing next generation lithium-ion batteries to meet, among other goals, energy storage, safety, and cost metrics to enable a plug-in hybrid electric car with an all-electric driving range of 40 miles to compete commercially with conventional gas-powered cars. In contrast, officials from ARPA-E reported that the BEEST initiative focused on developing ultra-high energy density, low-cost, transformational battery technologies beyond next generation lithium-ion batteries—such as metal-air and lithium-metal batteries—to enable an electric car that can travel 100 miles or more without needing to recharge to compete commercially with conventional gas-powered cars.

- *Differences in types of recipients.* Broadly overlapping initiatives reported differences in the types of recipients that were eligible to receive funding assistance. For example, both DOE's Batteries and Energy Storage Energy Innovation Hub and NSF's Energy for Sustainability Program reported that universities and industry were eligible to receive funding assistance. However, DOE officials reported that the hub had a broader number of eligible types of recipients. Specifically, DOE will award the hub funding to a lead entity that may be a DOE national laboratory, industry, university, or not-for-profit research institution. In addition, DOE officials also reported that DOD and other federal laboratories may participate in the hub despite not being eligible to lead it. The lead entity will be responsible for involving partners based in universities, private industry, nonprofits, and DOE's national laboratories and ensuring the hub has a broad focus on cross-disciplinary energy science, engineering, economics, and public policy. According to NSF officials, the agency typically supports single investigator research projects and makes its awards to university-based researchers who may receive additional support from industry partners in the form of funding or in-kind assistance, or even internship opportunities if the researcher is a graduate student.

While we did not find clear instances of duplicative initiatives, it is possible that there are duplicative activities among the initiatives that could be consolidated or resolved through enhanced coordination across agencies and at the initiative level. Also, it is possible that there are instances in which recipients receive funding from more than one federal source or that initiatives may fund some activities that would have otherwise sought and received private funding. Because it was beyond the scope of this work to look at the vast number of activities and individual awards that are encompassed in the initiatives we evaluated, we were unable to rule out the existence of any such duplication of activities or funding. Agency officials from most initiatives (30 of the 39) did report in survey responses that they include questions on their funding applications about other sources of federal funding that an applicant may be receiving for the same project. Officials from several of these agencies reported that they use this information to help avoid multiple funding of the same activity. For example, officials from NSF reported that potential awardees are required to provide detailed descriptions of their related work supported by federal funds. The scope and budget of the NSF award is adjusted if necessary to avoid multiple funding for any activity.

Agencies Reported Coordinating Their Initiatives with Each Other and Internally through Several Activities

Agencies reported several activities to coordinate with each other on their battery and energy storage initiatives, including nearly all initiatives that we identified as overlapping. In addition, DOE has taken steps to internally coordinate its electric vehicle battery and electric grid storage initiatives through several activities that, among other things, involved defining common technology goals. DOD has taken several actions to improve its coordination of battery and energy storage initiatives based on a recommendation in one of our prior reports. Agencies' actions were consistent with key practices we have previously identified that can help enhance and sustain federal agency coordination.[26] As we have

---

[26]GAO-06-15.

previously reported, a lack of coordination can waste scarce funds and limit the overall effectiveness of the federal effort.

Agencies Coordinated with Each Other through Several Activities

All six agencies reported coordinating with other each other on their battery and energy storage initiatives through a variety of activities. In addition, agency officials from nearly all initiatives that we identified as having overlapping characteristics (in that they supported similar technology advancement activities, technologies, uses, and goals) reported coordinating with other battery and energy storage initiatives. We have previously reported that coordination across programs may help address fragmentation, overlap, and duplication.[27] We grouped the various coordination activities that agency officials reported into five categories on the basis of our analysis. We found agency activities that were consistent with key practices we have previously identified that can help enhance and sustain federal agency coordination, including monitoring, evaluating, and reporting on results and agreeing on roles and responsibilities.[28] The categories of actions taken by agencies and examples of specific activities were as follows:

- *Participating in working groups.* Agency officials reported participating in several working groups. In particular, agency officials from DOE, DOD, and NASA identified the Chemical Working Group (CWG) of the Interagency Advanced Power Group as a key means for sharing information on energy storage science and technology advancement. The CWG provides an annual forum for agency program officials to present information on the progress and results of their ongoing projects to each other. The main participants in the CWG are DOE's ARPA-E, BES, Hydrogen Fuel Cells Program Office, and OVT; DOD's Air Force, Army, Defense Advanced Research Projects Agency (DARPA), and Navy; and NASA. Several of these offices implemented initiatives we identified previously as overlapping, such as ARPA-E's GRIDS initiative and the Navy and Marine Corps Energy Storage Science and Technology initiative. In December 2010, we reported that the CWG meetings have been effective in identifying instances of project duplication in the past.[29] According to the chairs of the CWG,

---

[27]GAO-11-318SP and GAO-12-677.

[28]GAO-06-15.

[29]GAO-11-113.

these meetings also create an opportunity for information exchange among agencies or their offices that has led to technology transfer. For example, these officials said that interactions fostered by the working group led to the transfer of knowledge developed by NASA on zinc-air battery technology to the Army and Marine Corps for use in their own projects. In addition to the CWG, agencies identified a number of other working groups, such as the NASA Power Steering Committee, the NASA Battery Working Group, the DOD Power Sources Technical Working Group, and the DOD Energy and Power Community of Interest. We have previously reported that federal agencies engaged in collaborative efforts should report on their activities to help key decision makers within the agencies, as well as clients and stakeholders, to obtain feedback for improving both policy and operational effectiveness.

- *Implementing interagency memorandums.* Agency officials identified several interagency memorandums. As a case in point, DOE and DOD signed a memorandum of understanding (MOU) in July 2010 to establish a framework for cooperation and partnership on energy issues. Under this MOU, DOE and DOD offices have implemented several joint programs and projects. For example, DOE's OVT and the Army's TARDEC established the Advanced Vehicle Power Technology Alliance (AVPTA) to coordinate their ground vehicle power research, including research on batteries for hybrid electric propulsion systems. In addition, ARPA-E and a number of DOD offices, including the Army Research Laboratory, TARDEC, Air Force Research Laboratory, Office of Naval Research, and Naval Surface Warfare Centers, as well as the Office of the Assistant Secretary of Defense for Operational Energy Plans and Programs (OEPP) and the Office of the Assistant Secretary of Defense for Research and Engineering established the Hybrid Energy Storage Module Integrated Project Team to develop coordinated funding announcements to support research on a variety of energy storage technologies for military and civilian uses. Further, DOE's OE Energy Storage Program and the Army's Ft. Leonard Wood jointly funded a microgrid energy storage demonstration project. Several of these DOE and DOD offices were involved in implementing overlapping initiatives that we identified. For example, we identified DOE's OVT Vehicle Technologies Energy Storage Research and Development initiative as having overlapping characteristics with DOD's Army Energy Storage for Air and Ground Vehicles initiative, which includes energy storage research and development activities supported by TARDEC. We have previously reported that collaborating agencies should work together to define and agree on their respective roles and responsibilities. In doing so, agencies can clarify which of them does what and organize their joint and individual efforts.

Coordinating directly through agency staff. Agencies also reported a number of activities that involved agency staff coordinating directly across initiatives. For example, DOE officials from BES reported that they engage in program manager discussions with other agencies, such as NSF and DOD, to help monitor their initiatives' activities and provide clear distinctions between DOE supported research and that funded by other agencies. As we noted previously, DOE's BES Batteries and Energy Storage Energy Innovation Hub had overlapping characteristics with NSF's Energy for Sustainability Program. In addition, Air Force officials from the Batteries for Space-Based Vehicles initiative, which primarily supports batteries for satellites, reported that they have conducted joint technology planning activities for space power and energy storage research needs with DOD's National Reconnaissance Office, which operates and maintains U.S. intelligence satellites. Agency officials also reported assisting each other with project selection and review to, among other things, reduce the risk of potential duplication of projects. For example, DARPA officials reported that officials from ARPA-E participated in DARPA's project selection panels for its Revolutionary Portable Energy Storage for the Warfighter initiative, and NASA officials with the Space Power Systems Project reported that they participate in DOE OVT's annual merit review and peer evaluation meetings. We have reported that federal agencies engaged in collaborative efforts need to create the means to monitor and evaluate their efforts to enable them to identify areas for improvement.

- *Sponsoring and participating in conferences and workshops.* Agencies reported sponsoring and participating in conferences and workshops with other agencies. For example, officials from DOE's BES reported that they sponsored a workshop that included officials from other agencies and that aimed to define common outcomes for DOE's basic research needs for electrical energy storage. Likewise, officials from the Air Force Special Purpose Power initiative reported participating in DOD's biannual Power Sources Conference, which coordinates current developments in power sources and energy storage development and allows the opportunity to report on the results of initiatives. We have previously reported that collaboration can help agencies to define and articulate the common federal outcome.

- *Checking applications for potentially duplicative federal funding.* As we mentioned previously, officials from most initiatives reported that they include questions on their funding applications about other sources of federal funding that an applicant may be receiving for the same project. Officials from several of these agencies reported that they use this information to help avoid multiple funding of the same activity.

DOE Has Taken Steps to Coordinate Its Electric Vehicle Battery and Grid Storage Initiatives

DOE has taken steps to internally coordinate its electric vehicle battery and electric grid storage initiatives through several actions. We found that these involved defining common technology goals; establishing strategies; and monitoring, evaluating, and reporting results. Specific steps are as follows:

- In September 2011, DOE completed an agencywide technology research and development plan, called its Quadrennial Technology Review (QTR), to guide its energy technology programs through 2015 under six strategies.[30] The QTR identified batteries and other energy storage technologies as critical elements in two of the strategies: vehicle electrification and electric grid modernization. DOE's vehicle electrification strategy is focused on developing technologies for light-duty electric vehicles to help significantly reduce oil consumption. These technologies include advanced rechargeable batteries and other components of electric vehicle systems.[31] DOE's grid modernization strategy is focused on developing technologies to help maintain reliable and secure electricity infrastructure that can meet expected growth in the demand for electricity in the future and integrate renewable energy technologies. In addition to energy storage, grid modernization includes advanced modeling of power distribution needs, and "smart grid" technologies that use computer-based remote control and automation devices such as sensors to improve monitoring and control of power flows.

---

[30]Department of Energy, *Report on the First Quadrennial Technology Review* (Washington, D.C.: September 2011).

[31]Other components of electric vehicle systems include, for example, electric power train and power electronics.

- DOE has also recently established two working groups, called integrated technical teams—one focused on electric vehicle batteries and the other on grid modernization. The technical teams provide staff from different DOE offices a mechanism to help monitor, evaluate, and report on their activities and results. For example, through the technical teams, participating offices review their funding announcements and conduct portfolio progress reviews of each other's ongoing work to avoid duplication between their initiatives. In addition, these teams have worked together to define common technology goals for DOE's battery and energy storage initiatives.

  - *Electric Vehicle Battery Technical Team.* This team includes officials from OVT, ARPA-E, and BES. The technical team set three common goals for DOE's electric vehicle battery technology advancement efforts: (1) the battery should not cost more than $0.01 per mile driven over the life of the battery; (2) the battery should get at least 10 miles of range per minute of charging time; and (3) should be safe, constructed of earth-abundant materials, and be recyclable.
  - *Grid Modernization Technical Team.* This team includes officials from the OE Energy Storage Program, ARPA-E, and BES. Among other actions, the offices on the team established joint strategies. Specifically, they cosponsored strategic planning workshops to determine agencywide priorities for applied research on batteries and other energy storage technologies for the electric grid, as well as specific performance targets for these technologies.[32] DOE has also established a goal of reducing the cost of energy storage on the grid by 30 percent by 2015, which the technical team is working together to achieve.

In addition to the technical teams, DOE offices reported other activities to coordinate directly with each other to help avoid duplicating efforts. For example, ARPA-E has formed a Panel of Senior Technical Advisors (PASTA) to coordinate with other DOE offices to leverage resources and ensure that ARPA-E provides unique value. As we reported in January 2012, PASTA is a group of DOE managers that meet periodically to

---

[32]Sarah Lichtner, Ross Brindle, and Lindsay Pack, *Electric Power Industry Needs for Grid-Scale Storage Applications,* special report prepared at the request of the Department of Energy, December 2010.

discuss current and future DOE research efforts.[33] PASTA includes managers from ARPA-E and, among others, the Office of Science, EERE, and OE.

DOD Has Taken Steps to Improve Its Coordination of Battery and Energy Storage Initiatives

DOD officials told us that DOD has taken the following actions to improve its coordination of battery and energy storage initiatives, both with other agencies and, internally, among its offices and the services:

- A DOD official from OEPP told us that DOD has taken two actions to improve coordination with other agencies on battery and energy storage initiatives in response to a past GAO recommendation. In 2010, we reported that DOD's coordination on power sources science and technology, which includes batteries and other energy storage technologies, was generally effective. However, we concluded that the agency may be missing opportunities to leverage resources and avoid offices initiating similar research projects because participation in some coordination activities was voluntary and could be more complete.[34] We recommended that DOD determine methods to strengthen participation in interagency coordination mechanisms. To address our recommendation:

  - A DOD official from OEPP told us that DOD's Assistant Secretaries of Defense for Research and Engineering and for OEPP have assigned the DOD Energy and Power Community of Interest (EPCOI) the responsibility for ensuring interagency coordination mechanisms on power sources science and technology. The EPCOI working group involves officials from the Army, Air Force, and Navy to coordinate DOD energy and power science and technology activities, including batteries and other energy storage technologies.
  - As described earlier in this report, DOD and DOE offices have implemented several joint programs and projects as a result of the July 2010 MOU between the agencies to coordinate on energy issues.

---

[33]GAO, *Department of Energy: Advanced Research Projects Agency-Energy Could Benefit from Information on Applicants' Prior Funding,* GAO-12-112 (Washington, D.C.: Jan. 13, 2012).

[34]GAO-11-113.

- DOD officials told us that DOD has taken two actions to improve its coordination of battery and energy storage initiatives among its offices and the services. First, officials from DOD's OEPP told us that DOD has created an operational energy[35] strategy that DOD issued in May 2011, as well as an implementation plan for this strategy that DOD issued in March 2012. These plans established departmentwide priorities for reducing demand for operational energy through, in part, investment in new technologies and equipment such as lighter batteries.[36] DOD officials told us that the strategy will help coordinate DOD's battery and energy storage initiatives because it provides an agencywide road map for incorporating operational energy priorities, such as energy storage science and technology investments, in DOD's programs. Second, DOD has also tasked OEPP to certify that the budget submissions of DOD offices and the services reflect these priorities.

Agency Comments and Our Evaluation

We provided a draft of this report for review and comment to DOE, DOD, NASA, NSF, EPA, and NIST. DOE, DOD, NASA, and NSF provided technical and clarifying comments, which we incorporated as appropriate. The other agencies we reviewed had no comments.

As agreed with your offices, unless you publicly announce the contents of this report earlier, we plan no further distribution until 30 days from the report date. At that time, we will send copies to the Secretaries of Defense, Energy, and Commerce; the Administrators of the EPA and NASA; the Directors of NIST and NSF; the appropriate congressional committees; and other interested parties. In addition, the report will be available at no charge on the GAO website at http://www.gao.gov.

---

[35]Operational energy is defined as the energy required for training, moving, and sustaining military forces and weapons platforms for military operations. The term includes energy used by tactical power systems and generators and weapons platforms.

[36]Department of Defense, *Energy for the Warfighter: Operational Energy Strategy*, (Washington, D.C.: May 2011) and *Operational Energy Strategy: Implementation Plan* (Washington, D.C.: March 2012).

GAO-12-842 Batteries and Energy Storage

If you or your staff have any questions about this report, please contact me at (202) 512-3841 or ruscof@gao.gov. Contact points for our Offices of Congressional Relations and Public Affairs may be found on the last page of this report. GAO staff who made key contributions to this report are listed in appendix VI.

Frank Rusco
Director
Natural Resources and Environment

Appendix I: Objectives, Scope, and Methodology

Our objectives were to (1) identify the scope and key characteristics of federal battery and energy storage initiatives; (2) determine the extent to which there is potential fragmentation, overlap, or duplication of these initiatives, if any; and (3) determine the extent to which agencies coordinate their battery and energy storage initiatives.

To inform our overall review and help us understand the range of batteries and other energy storage technologies, their uses, and the increased interest in these technologies in recent years, we collected background information from several sources. Specifically, we reviewed a 2010 GAO report that identified Department of Defense (DOD) investments in power sources, including batteries and other energy storage technologies.[1] We also consulted analysts from the Congressional Research Service (CRS) and the Department of Energy's (DOE) Office of Inspector General who have subject-matter expertise on federal battery and energy storage initiatives. We also interviewed officials from DOE's Advanced Research Projects Agency-Energy (ARPA-E), Loan Programs Office, Office of Electricity Delivery and Energy Reliability, Office of Energy Efficiency and Renewable Energy's Office of Vehicle Technologies (OVT), and Office of Science's Office of Basic Energy Sciences to learn about their ongoing and planned initiatives, as well as officials from the Federal Energy Regulatory Commission's Office of Energy Policy and Innovation regarding recent electricity market regulatory developments relevant to batteries and other energy storage technologies.

To identify the scope and key characteristics of federal battery and energy storage initiatives, we first decided to focus our review on rechargeable (i.e., secondary) batteries and other technologies but excluded nonrechargeable (i.e., primary) batteries, fuel cells, and nuclear energy storage technologies. In addition, we focused on initiatives that were active[2] in any year during fiscal years 2009 through 2012 because, during these years, DOE offices made substantial investments in these technologies, including with funds made available under the 2009 American Recovery and Reinvestment Act (Recovery Act). For example, DOE's ARPA-E supported about $97 million in battery and energy

---

[1]GAO-11-113.

[2]For the purposes of this report, we defined active initiatives as those that were planned or funded or implemented or authorized in any of the fiscal years described.

storage technologies during fiscal years 2010 and 2011 with funding
made available under the Recovery Act. We then developed several
definitions to help us identify initiatives. For example, we defined an
initiative as a group of agency activities serving a similar purpose or
function, such as a program or mission area. For the purposes of the
review, we also developed definitions for technology advancement
activities: basic science, applied research, demonstrations,
commercialization, and deployment activities.[3] We limited our review to
initiatives that supported basic science, applied research, demonstrations,
and commercialization activities for batteries and other energy storage
technologies. Because initiatives often supported more than one type of
technology advancement activity, some of the initiatives included in this
report may also support deployment activities. However, we excluded any
initiatives that focused solely on deployment and initiatives that involved
agency-owned assets such as fleets or facilities. For example, we did not
include the Department of the Treasury (Treasury) and several federal tax
credits it administered that indirectly supported deployment of battery and
energy storage technologies during fiscal years 2009 through 2012.
Treasury officials told us data were generally not available on the
estimated revenue loss directly associated with batteries and other
energy storage technologies because available data do not break out
qualifying investments in fine enough detail to determine which part may
have been for qualifying battery and energy storage devices.

We next took steps to identify the agencies and initiatives that supported
these technologies and were in our scope. To do this, we first compiled a
preliminary list of agencies and initiatives based on two previous GAO
reports that collected information on some federal battery and energy

---

[3]We developed definitions that could be applied broadly to make comparisons across
agencies and that covered the full spectrum of advancement activities. Federal agencies
use various definitions and categories for describing the stages of technology
advancement. For example, the National Aeronautics and Space Administration (NASA)
and DOE use technology readiness level (TRL) categories and definitions to measure and
communicate technology readiness for first-of-a-kind technology applications. However,
these agencies' TRL categories and definitions are not the same. In addition, the Office of
Management and Budget (OMB) provides federal definitions for the terms basic research,
applied research, and development in OMB Circular No. A–11, Section 84—Character
Classification (Schedule C) but does not provide definitions for demonstrations and
commercialization activities. During pretests of our questionnaire, agency officials were
able to fit their initiatives' activities within our categories and based on our definitions,
which validated our use of them.

storage initiatives active in fiscal year 2010.[4] In addition, as mentioned previously, we interviewed DOE officials and also reviewed DOE documents. We also searched several databases and websites, including the Catalog of Federal Domestic Assistance, National Technical Information Service, Article First, ECO, Worldcat, and the Congressional Research Service for materials published in the last 10 years that might identify relevant agencies and initiatives. We shared a preliminary list of initiatives with agency officials, along with information about our scope and definitions for key terms—such as batteries and energy storage technologies, active initiatives, and technology advancement activities— and asked officials to confirm the initiatives on our list and also which initiatives should be added. If agency officials indicated they wanted to remove an initiative from or consolidate initiatives on our list, we asked for additional information. For example, we removed from our list an initiative in the Department of Transportation's National Highway Traffic Safety Administration that we learned focused on developing safety measures regarding the batteries in commercially available electric vehicles and not on technology advancement activities. In addition, DOD officials told us they wanted to consolidate their programs and projects into distinct initiatives for each of the military services for the purpose of our review because their programs and projects cut across multiple offices. Through interviews and other correspondence with agency officials, we confirmed a final list of six agencies and 39 federal battery and energy storage initiatives. We may not have identified and included every battery and energy storage initiative in the agencies we reviewed; however, given our methodology, we believe we have identified most, if not all.

We then developed a questionnaire to collect information from officials involved in the 39 initiatives we identified at the six agencies on the initiatives' key characteristics. We asked each agency to confirm the names and contact information for the officials most knowledgeable about each initiative, and thus the most appropriate person to complete the questionnaire. Specifically, we developed questions that asked about the types of technology advancement activities supported, technologies and uses supported, whether staff conducted in-house activities or provided assistance to external recipients, types of eligible recipients, and types of funding mechanisms used to provide assistance. We also asked about the initiative's mission and overarching goals, its goals, if any, directly

---

[4]GAO-11-113 and GAO-12-260.

related to battery and energy storage technologies, performance measures, as well as technical cost and technical performance targets that the initiative may have. We asked about obligations for fiscal years 2009 through 2012. Finally, we also asked whether staff from the initiative formally coordinated with other battery and energy storage initiatives within their agency or with other federal agencies and to describe examples of any coordination activities. To minimize errors arising from the differences in how questions might be interpreted and to reduce variability in responses that should be qualitatively the same, we worked with an independent GAO questionnaire specialist and conducted pretests of draft questionnaires with officials we had identified from four different initiatives at four different agencies during February and March 2012. We conducted the pretests to check that (1) the questions were clear and unambiguous, (2) terminology was used correctly, (3) the questionnaire did not place an undue burden on agency officials, (4) the information could feasibly be obtained, and (5) the questionnaire was comprehensive and unbiased. A second independent GAO questionnaire specialist also reviewed a draft of the questionnaire prior to its administration. On the basis of feedback from these pretests and independent review, we revised the questionnaire in order to improve its clarity.[5]

After completing the pretests, we administered the questionnaire. We sent questionnaires to the appropriate agency officials in an attached Microsoft Word form. We received questionnaire responses for each initiative and, thus, had a response rate of 100 percent. After analyzing the responses, we conducted follow-up e-mail exchanges or telephone discussions with agency officials when responses were unclear or conflicting, such as when both "Yes" and "No" boxes were checked or boxes were left completely unchecked. When necessary, we used the clarifying information provided by agency officials to update answers to questions to improve the accuracy and completeness of the data. Some officials reported that the funding obligations data had to be estimated because the initiative did not track its obligations for battery and energy storage-specific activities. In addition, all funding obligations reported for

---

[5]For example, to help agency officials precisely understand our technology scope, we explained in the introduction to our questionnaire that our definition of batteries and other energy storage technologies encompassed full systems, as well as subsystems, components, and power management approaches for these technologies—for example, battery components, battery cells and packs, and battery power management systems.

fiscal year 2012 were estimated. For these cases, we conducted follow-up interviews and reviewed relevant supplemental documentation to understand how the officials arrived at estimates and to assess the reliability of these estimates for our purposes. Overall, we determined agencies' estimates to be sufficiently reliable for purposes of our report. As a result of our follow-up discussions with agency officials to clarify responses, we determined in consultation with the agencies that one DOD and one NASA initiative did not meet our original definitions of a battery and energy storage initiative. Thus, we removed these initiatives from our final list, which changed from 41 to 39 initiatives. According to an official from DOE's OVT, since the survey focused on research and development activities, they did not include in their survey response $1.5 billion in funds made available under the Recovery Act. With these funds DOE made awards to 20 projects to support the establishment of advanced battery manufacturing and battery recycling facilities in the U.S. DOE officials provided this information separately in OVT's annual energy storage research progress reports.[6] A copy of our questionnaire is presented in appendix V.

We used standard descriptive statistics to analyze responses to the questionnaire. Because this effort was not a sample questionnaire, it has no sampling errors. However, the practical difficulties of conducting any questionnaire may introduce errors, commonly referred to as nonsampling errors. For example, difficulties in interpreting a particular question, sources of information available to respondents, or entering data into a database or analyzing them can introduce unwanted variability into the questionnaire results. We took steps in developing the questionnaire and collecting and analyzing the data to minimize such nonsampling errors. For example, social science questionnaire specialists designed the questionnaire in collaboration with GAO staff that had subject-matter expertise. As previously mentioned, we pretested the draft questionnaire to ensure that the questions were clearly stated and easy to understand. When we analyzed the data using computer programs, an independent analyst checked the results from all the computer programs. Finally, we verified the accuracy of a small sample of keypunched records by comparing them with their corresponding questionnaires, and we corrected the errors we found. Less than 0.5 percent of the data items we

---

[6]See, for example, Department of Energy, Energy Efficiency and Renewable Energy, *Fiscal Year 2011 Annual Progress Report for Energy Storage R&D*, DOE/EE-0675 (Washington, D.C.: Jan. 2012).

checked had random keypunch errors that would not have been corrected
during data processing. While we did not verify all responses, on the
basis of our application of recognized questionnaire design practices and
follow-up procedures, we determined that the data used in this report
were of sufficient quality for our purposes.

To determine the extent to which there is potential fragmentation, overlap,
or duplication of these initiatives, if any, we developed definitions for
these terms based on definitions established in prior GAO reports.
Specifically, for the purposes of this report, fragmentation, overlap, and
duplication, were defined as follows:

- Fragmentation occurs when more than one federal agency (or more
  than one organization within an agency) is involved in the same broad
  area of national need.

- Overlap occurs when multiple initiatives support similar technologies,
  uses, technology advancement activities, and funding recipients, and
  have similar goals.

- Duplication occurs when multiple initiatives support the same
  technology advancement activities for the same technologies and
  uses, provide funding to the same recipients using the same funding
  mechanisms, and have the same goals.

To analyze potential fragmentation, we used agencies' questionnaire
responses to confirm the number of federal agencies that supported
battery and energy storage initiatives. We compared this information with
our GAO definition for fragmentation. To analyze potential overlap and
duplication, we analyzed the key characteristics provided by agencies for
each initiative in their questionnaire responses. Specifically, we compared
each initiative with each of the other 38 initiatives to determine whether
agency officials reported that the initiatives supported at least one similar
technology advancement activity, type of battery and other energy
storage technology, and category of use. If the initiatives did not share
one of these three characteristics, they were considered not to have
potential overlap. For two initiatives that reported similar responses for all
three characteristics, we compared additional information from
questionnaire responses regarding the initiatives' missions, goals,
performance measures, technical cost metrics and technical performance
metrics (if any were reported), to determine whether the initiatives also
had similar missions or goals. If so, we determined the initiatives were
potentially overlapping. To complete the analysis, two GAO

scientists/engineers independently performed these comparisons and
compared their findings. They discussed their independent findings to
come to a joint decision as to whether initiatives were potentially
overlapping or not. If an initiative was determined to be potentially
overlapping, then we considered it to hold potential for duplication as well.
For initiatives that we identified as potentially overlapping, we held
interviews with agency officials to gather more information to verify that
they were in fact similar. We did not interview officials involved in every
initiative that we identified as potentially overlapping but focused on
selected ones in DOE, DOD, and the National Science Foundation (NSF)
primarily. We also reviewed questionnaire responses regarding each
battery and energy storage initiative's mission and overarching goals to
identify how the initiative supported the agency's overall mission.

To determine the extent to which agencies coordinate their initiatives, we
used questionnaire responses and interviews to identify interagency
coordination activities across the six agencies. For determining the extent
of internal coordination, we examined internal coordination in DOE and
DOD of initiatives within those agencies because they had the largest
number of initiatives and amount of funding obligations among the
agencies we reviewed. We also followed up with DOD on actions, if any,
it took to address recommendations we made in 2010. We did not assess
internal coordination in NASA, NSF, the Environmental Protection Agency
(EPA), or the National Institute of Standards and Technology (NIST)
because total funding obligations for these initiatives were substantially
smaller than for DOE and DOD. For example, NASA's total obligations
were about $21 million, and DOE's were about $852 million for fiscal
years 2009 through 2012. For the purposes of this report, we defined
coordination as any joint activity by two or more organizations that is
intended to produce more public value than could be produced when the
organizations act alone. We drew on past GAO work related to
interagency coordination to help us identify agency coordination
activities.[7]

We conducted this performance audit from September 2011 to August
2012 in accordance with generally accepted government auditing
standards. Those standards require that we plan and perform the audit to
obtain sufficient, appropriate evidence to provide a reasonable basis for

---

[7]GAO-06-15.

our findings and conclusions based on our audit objectives. We believe
that the evidence obtained provides a reasonable basis for our findings
and conclusions based on our audit objectives.

Appendix II: Batteries and Other Energy Storage Technologies

Agencies reported supporting a number of batteries and other energy storage technologies through their initiatives. Table 7 provides descriptions of technologies identified during our review.

Table 7: Batteries and Other Energy Storage Technologies

Rechargeable batteries	
Advanced lead-acid batteries	These are improved versions of the 100 year old battery used to provide power for starting vehicle engines. These batteries use lead as the anode, lead dioxide as the cathode, and a sulfuric acid electrolyte. Advanced lead-acid batteries are considered suitable for stationary storage uses. They are commercially available but also are being researched to, among other goals, improve the amount of time they can be usefully discharged and recharged.
Flow batteries[a]	Flow batteries—such as vanadium redox and zinc-bromide—store energy in electrolyte solutions that are contained in external tanks. They are candidates for use in stationary storage systems. Some flow batteries are commercially available or being demonstrated. Most types are also being researched.
Lithium-ion batteries	These are batteries in which lithium ions move from the cathode to the anode during the discharging and charging processes. These batteries are the most popular type of rechargeable battery for use in personal electronics and increasingly for electric vehicles and stationary storage systems. Research efforts aim to improve their energy capacity, safety, and reduce their cost.
Lithium-metal batteries	These are batteries that use lithium as the anode. They hold the potential for providing greater energy stored per unit weight compared with lithium-ion batteries. These batteries are currently being researched primarily for electric vehicles, although they have other potential uses. Research aims to, among other goals, develop new materials and battery cell designs.
Metal-air batteries	These are batteries that use oxygen as the cathode and a metal anode such as magnesium, iron, or lithium. They hold potential for providing higher energy density and lower costs than lithium-ion batteries. These batteries are currently being researched for electric vehicles and stationary storage uses. Research aims to, among other goals, improve their ability to recharge.
Sodium batteries	Batteries that use sodium, or sodium compounds as electrodes. They are primarily considered suited for use in stationary storage systems. Some sodium batteries are commercially available, but others are being researched. Currently, research aims to, among other goals, develop new materials and battery cell designs.
Other energy storage technologies	
Capacitors	These are devices that store energy in an electrostatic charge that can withstand hundreds of thousands of charge and discharge cycles without degrading. Capacitors have been used for small, primarily consumer electronic devices and are increasingly being developed for high-power weaponry and commercial electric vehicles. Currently, research aims to, among other goals, increase their energy density.
Compressed air energy storage (CAES)	This is a storage system that involves injecting compressed air into a geological formation such as an underground cavern within a salt dome. To recover the power, the air is released and used to help drive a turbine generator. CAES provides bulk energy storage for stationary power systems. Currently, CAES is a mature technology; however, research exists to, among other things, develop new turbine technologies.
Flywheels	These are devices that store electricity in the form of mechanical energy in a spinning wheel or tube. To recover power, the flywheel drives a generator. Flywheels provide high power and quick release of energy over short durations. Flywheels are commercially available; however, research is being done to, among other goals, find new and improved device materials.

Hydraulic accumulator energy storage[b]	These are devices that store energy in the form of pressurized pumped fluid. Accumulators are being demonstrated and researched for use in hybrid vehicles that combine a hydraulic pump-powered motor with a gas-powered motor to provide vehicle propulsion.
Solar thermal energy storage[b]	These are systems that store heat from solar energy collection in a medium such as molten salt and release it to drive a turbine electricity generator. The technology is currently commercial in small amounts and limited by the need for suitable geography. There is ongoing research to develop thermal energy storage materials and storage methods.
Superconducting magnetic energy storage[b]	These are devices that store energy in the form of a magnetic field. These devices are used for stationary storage systems because they provide short bursts of energy very rapidly to help support power grid electricity reliability. They are commercially available but are also being researched to develop systems with lower costs.

Sources: GAO analysis of CRS, DOE, and EPA documents.

[a]Agency officials reported that their initiatives supported flow batteries other than redox flow batteries by using the "other batteries" field provided in our questionnaire. See appendix IV for selected questionnaire responses for the initiatives we reviewed. We reported initiatives' support for these batteries under "other" technologies (column 11).

[b]Agency officials reported that their initiative supported this type of energy storage technology using the "other technologies" field provided in our questionnaire. See appendix IV for selected questionnaire responses for the initiatives we reviewed. We reported initiatives' support for this technology under "other" technologies (column 11).

Appendix III: Federal Battery and Energy Storage Initiatives

We identified 39 initiatives across six agencies that supported batteries and other energy storage technologies through basic research, applied research, demonstrations, commercialization, and deployment activities. Tables 8, 9, 10, 11, 12, and 13 provide descriptions of the initiatives in each agency by implementing office.

Table 8: Department of Energy Battery and Energy Storage Initiatives and Their Total Funding Obligations for Fiscal Years 2009 through 2012

Name of initiative and description	Subtotal (actual and estimated) obligations fiscal years 2009 through 2012
Vehicle Technologies Energy Storage Research and Development: This initiative, implemented by the Office of Energy Efficiency and Renewable Energy's Office of Vehicle Technologies (OVT), advances battery technologies to enable a large market penetration of hybrid and electric vehicles through applied materials research and battery development activities with national labs, universities, and industry.	$323,043,000[a]
Office of Electricity Energy Storage Program: This program, implemented by the Office of Electricity Delivery and Energy Reliability (OE), is designed to develop and demonstrate new and advanced energy storage technologies that will enable the stability and surety of the future electric utility grid as it transforms into a smart grid. The program focuses on technology development that improves the affordability and performance of energy storage and enables a robust suite of competitive options for various grid services.	241,234,515
Core Research - Materials Science and Engineering: This initiative, implemented by the Office of Science's Office of Basic Energy Sciences (BES), encompasses research activities that support basic science research to provide the knowledge base for the discovery and design of new materials for the generation, storage, and use of energy and for mitigation of the environmental impacts of energy use.	23,170,874
Energy Frontier Research Centers: This initiative, implemented by BES, supports basic and advanced discovery research to accelerate advanced energy technologies, including renewable energy technologies, by combining the talents and creativity of the nation's scientific workforce with a powerful new generation of tools for penetrating, understanding, and manipulating matter on the atomic and molecular scales.	69,500,000
Batteries and Energy Storage Energy Innovation Hub: This initiative, implemented by BES, is focused on transforming electrochemical energy storage beyond the current limits, including the exploration of new materials, architectures, systems, and novel approaches for transportation and utility-scale storage, by supporting cross-disciplinary research and development.	20,000,000
Batteries for Electrical Energy Storage in Transportation[b]: This initiative, implemented by the Advanced Research Projects Agency-Energy (ARPA-E), seeks to develop a new generation of ultra-high energy density, low-cost battery technologies for long electric range plug-in hybrid electric vehicles and all-electric vehicles.	36,344,516
Grid-Scale Rampable Intermittent Dispatchable Storage[b]: This initiative, implemented by ARPA-E, seeks to develop new technologies to enable the widespread deployment of cost-effective, grid-scale energy storage technologies to balance the short-duration variability in renewable generation.	27,687,068
ARPA-E Initial Funding Opportunity Announcement—Energy Storage Technologies[b]: This initiative was ARPA-E's first funding announcement and primarily aimed at prospective applicants who already had a relatively well-formed research and development plan for a transformational concept or new technology, including energy storage technologies, that could make a significant contribution toward attainment of the administration's Energy and Environment Agenda, if and when successfully deployed.	33,052,915

Name of initiative and description	Subtotal (actual and estimated) obligations fiscal years 2009 through 2012
High Energy Advanced Thermal Storage[b]: This initiative, implemented by ARPA-E, seeks to develop revolutionary cost-effective thermal energy storage technologies in three focus areas: (1) high-temperature storage systems to deliver solar electricity more efficiently around the clock and allow nuclear and fossil baseload resources the flexibility to meet peak demand; (2) fuel produced from the sun's heat; and (3) heating, ventilation, and air conditioning systems that use thermal storage to dramatically improve the driving range of electric vehicles.	37,268,245
Advanced Technology Vehicles Manufacturing Loan Program: This program, implemented by the Loan Programs Office (LPO), provides direct loans to companies making cars and components in U.S. factories that increase fuel economy at least 25 percent above 2005 fuel economy levels.	29,627,800[d]
Title XVII Loan Guarantee Program[c]: This program, implemented by LPO, provides loan guarantees, under Section 1703, to innovative clean technologies that avoid, reduce or sequester air pollutants. Under Section 1705 (now sunsetted), the program provided loan guarantees to certain clean energy projects, including those employing more mature technologies, that began construction prior to September 30, 2011.	11,065,875[d]
Total	**$851,994,808**

Source: GAO analysis of agency-provided data.

[a]DOE's OVT also awarded $1.5 billion in funds made available under the Recovery Act to 20 projects to support the establishment of advanced battery manufacturing and recycling facilities in the United States.

[b]ARPA-E's authorizing legislation, the America Creating Opportunities to Meaningfully Promote Excellence in Technology, Education, and Science Reauthorization Act of 2007, (Pub. L. No. 110-69 (2007)) was reauthorized on January 4, 2011, and funding for the program was authorized through fiscal year 2013.

[c]The Title XVII Loan Guarantee Program is in Title XVII of the Energy Policy Act of 2005, (Pub. L. No. 109-58 (2005)). The American Recovery and Reinvestment Act of 2009 added Section 1705 to Title XVII. The Section 1705 initiative had a sunset date of September 30, 2011.

[d]These obligations represent the credit subsidy costs associated with about $596 million in federal loans supported by DOE for battery and energy storage projects. Specifically, the Advanced Technology Vehicles Manufacturing Loan Program made about $520 million in loans and the Title XVII Loan Guarantee Program guaranteed about $76 million in loans. Loans supported by the Title XVII Loan Guarantee Program were made by the U.S. Treasury's Federal Financing Bank. Credit subsidy costs are the estimated net long-term cost to the government, in present value terms, of the loans over the entire period the loans are outstanding. Present value is the worth of the future stream of returns or costs in terms of money paid immediately. In calculating present value, prevailing interest rates provide the basis for converting future amounts into their "money now" equivalents. Credit subsidy costs exclude administrative costs and any incidental effects on governmental receipts or outlays. Officials from DOE's LPO told us that one of the battery and energy storage projects under the Title XVII Loan Guarantee Program with a loan guarantee commitment for a loan of $17 million withdrew from the program reducing the value of its loan to zero. As a result, the total amount in loans that DOE guaranteed for these technologies under the program was reduced by $17 million, and the total amount of the program's credit subsidy obligations were also reduced by over $830,000.

Table 9: Department of Defense Battery and Energy Storage Initiatives and Their Total Funding Obligations for Fiscal Years 2009 through 2012

Name of initiative and description	Subtotal (actual and estimated) obligations fiscal years 2009 through 2012
Army Energy Storage for the Soldier: This initiative, implemented by the Office of the Assistant Secretary of the Army for Acquisition, Logistics and Technology, aims to provide the warfighter with an ergonomic, lightweight, high energy density, soldier-worn battery that can serve as a central power source for all worn and carried power consumers.	$16,000,000[a]
Army Energy Storage for Basing: This initiative, implemented by the Office of the Assistant Secretary of the Army for Acquisition, Logistics and Technology, examines technologies to manage, distribute and store energy in bases of all sizes (from tactical contingency bases to large fixed installations). These sources require some level of energy storage, especially in the development of microgrids and other architectures for power management.	13,100,000[a]
Army Energy Storage for Air and Ground Vehicles: This initiative, implemented by the Office of the Assistant Secretary of the Army for Acquisition, Logistics and Technology, conducts research and development to enable hybridization, auxiliary power units, and other nonprime power applications.	80,058,000[a]
Navy and Marine Corps Energy Storage Science and Technology: This initiative, implemented by the Office of Naval Research, Sea Warfare Department (ONR 33), supports fundamental and early applied research to identify and develop novel materials and device architectures with improved power and energy densities over current state-of-the-art technology to address Navy/Marine Corps-unique requirements. The initiative supports applied research and prototype development to develop, demonstrate, and transition practical devices and systems that exploit novel energy storage materials, devices, and designs in practical embodiments to Navy/Marine Corps acquisition programs for further maturation and fielding.	80,091,000[a]
Navy and Marine Corps Energy Storage System Integration: This initiative, implemented by the Naval Sea Systems Command, Naval Air Systems Command, and U.S. Marine Corps, conducts applied research, demonstration, and deployment activities to enable and develop systems, technologies, processes, and materials to allow large format lithium-ion batteries to be carried, installed, charged, maintained, and utilized aboard surface ships and submarines, enable operational missions supporting the warfighter, and enhance mission coverage and capability.	130,000,000[a]
Air Force Basic Research Activities in Battery and Storage Technologies: This initiative, implemented by the Air Force Research Laboratory, Air Force Office of Scientific Research, conducts basic research programs focused on advancement and understanding of batteries and storage technologies.	9,173,000
Air Force Wide Temperature Capacitor Research and Development: This initiative, implemented by the Air Force Research Laboratory-Propulsion Directorate, focuses on the development of wide temperature capacitor dielectric materials, modeling and simulation of capacitor architectures and geometries, and the development of specialized testing capabilities for both dielectric materials and packaged capacitors.	4,515,000
Air Force Batteries for Aircraft and Directed Energy Weapons: This initiative, implemented by the Air Force Research Laboratory-Propulsion Directorate, under the Integrated Vehicle Energy Technology program, aims to develop electrochemical energy storage systems to provide safe, lightweight electrical power for aircraft and directed energy weapons.	24,436,646
Air Force Zinc-Bromide Flow Battery: This initiative, implemented by the Air Force Research Laboratory-Materials and Manufacturing Directorate, was funded only in fiscal year 2010 and supported testing and evaluation of zinc-bromide flow batteries for use when connected to a microgrid and solar generation system to support the military mission and provide power in the event of grid outage.	3,337,050

Name of initiative and description	Subtotal (actual and estimated) obligations fiscal years 2009 through 2012
Air Force Special Purpose Power: This initiative, implemented by the Air Force Research Laboratory-Propulsion Directorate, supports development of novel power systems and technologies for specialized Air Force applications. Specific applications include: airman portable power systems and unmanned aerial vehicle power/propulsion systems.	11,663,000
Air Force Advanced Materials for Energy Storage Applications: This initiative, implemented by the Air Force Research Laboratory-Materials and Manufacturing Directorate, supports fundamental, as well as applied research and development efforts to develop a variety of high-performance materials and manufacturing technologies necessary to increase the performance of next generation energy storage devices, such as batteries and capacitors, for a variety of military applications.	22,590,000[a]
Air Force Batteries for Space-Based Vehicles: This initiative, implemented by the Air Force Research Laboratory-Space Vehicles Directorate, aims to develop energy storage systems (batteries and ultracapacitors) to meet the needs of DOD space-based assets, to include high specific energy and power, long cycle life, and improved depth of discharge.	7,669,579
DARPA Revolutionary Portable Energy Storage for the Warfighter: This initiative, implemented by the Defense Advanced Research Projects Agency (DARPA), aimed to solve high-risk, DOD mission-critical portable power and energy challenges that are unique to DOD, not adequately addressed by the commercial market, and lie beyond the capabilities, scope, and risk level of other DOD organizations.	14,662,444
ESTCP Installation Energy Test Bed: This initiative, implemented by the Office of the Deputy Under Secretary of Defense for Installations and Environment, Environmental Security Technology Certification Program (ESTCP), demonstrates renewable energy, energy efficiency, energy management, and energy storage technologies. Energy storage technologies are demonstrated in the context of microgrids that will enable grid-compatible operation, improve efficiency of an installation's power network, and enable increased use of distributed generation, especially renewable energy sources.	12,978,510[a]
Total	**$430,274,229**

Source: GAO analysis of agency-provided data.

[a]All or part of the agency-provided obligations are estimated. For example, Army officials told us that it was hard to break out obligations for rechargeable batteries from obligations for all batteries—including nonrechargeable batteries— supported under its initiatives. Therefore, Army officials' reported obligations were an overestimate of the actual amount obligated for rechargeable batteries.

Table 10: National Aeronautics and Space Administration Battery and Energy Storage Initiatives and Their Total Funding Obligations for Fiscal Years 2009 through 2012

Name of initiative and description	Subtotal (actual and estimated) obligations fiscal years 2009 through 2012
Prognostics Algorithm Development: This initiative, implemented by Ames Research Center, investigates algorithms to estimate the state of health, and remaining life of components, and applies it to batteries, including lithium-ion batteries. The key goal of this initiative is to develop and mature the science of prognostics for systems health management in aerospace applications.	$500,000[a]
Night Rover Challenge: This initiative, implemented by the Office of the Chief Technologist's Centennial Challenges Program, is a $1.5 million prize purse competition to develop an energy storage system that can store at least 500 watt-hours per kilogram.	1,500,000
Silicon Nano-Wire Anode: This initiative, implemented by the Office of the Chief Technologist, seeks to adapt proven, breakthrough battery technology relying on nanostructured silicon anodes (silicon nanowire technology) to the extreme environments of space applications.	1,200,000
Space Power Systems Project: This initiative, implemented by the Office of the Chief Technologist, Game Changing Development Program, seeks to improve the performance of secondary lithium-ion battery cells to meet the energy storage requirements of human space missions by developing advanced battery components that safely provide substantially higher specific energy and energy density than is currently available.	13,422,620
Flywheel Energy Storage and Momentum Control: This initiative, implemented by Glenn Research Center, conducts applied research to develop and demonstrate unique NASA cross-cutting flywheel technology for space and terrestrial applications and commercialize the technology through direct partnership with industry and intellectual property licensing.	3,008,754[a]
Aerospace Lithium-Ion Cell Qualification Program: This initiative, implemented by Goddard Space Flight Center, performs cell characterization tests followed by simulated life cycling regimes at both low Earth orbit and geosynchronous Earth orbit of large prismatic lithium-ion cells from several vendors to qualify them for aerospace use.	80,000
Lithium-Ion COTS Battery Surveillance: This program, implemented by Johnson Space Center, involves a surveillance of state-of-the-art commercial-off-the-shelf (COTS) lithium-ion battery cells. Lithium-ion cells are purchased and undergo a stringent set of performance and safety tests to determine their suitability for space applications in a human-rated environment. The information collected enables quick design, buildup, and test at the battery level (as cell level data is already existent through this program). This initiative also included studies of polymer lithium-ion cells and in-depth study of issues encountered in the battery industry with respect to safety.	850,000
NASA Space Act Agreement with Underwriters Laboratories: This initiative, implemented by Johnson Space Center, aims to obtain a standard for simulating internal shorts in lithium-ion batteries for various applications.	250,000
Total	**$20,811,374**

Source: GAO analysis of agency-provided data.

[a] All or part of the agency-provided obligations are estimated.

Table 11: National Science Foundation Battery and Energy Storage Initiatives and Their Total Funding Obligations for Fiscal Years 2009 through 2012

Name of initiative and description	Subtotal (actual and estimated) obligations fiscal years 2009 through 2012
Energy for Sustainability: This initiative, implemented by the Directorate for Engineering, supports fundamental research and education that will enable innovative processes for the sustainable production of electricity and transportation fuels. One research interest area is on high-energy density and high-power density batteries suitable for transportation applications.	Information not available[a]
Energy, Power, and Adaptive Systems: This initiative, implemented by the Directorate for Engineering, supports the design and study of intelligent and adaptive engineering networks with an emphasis on electric power electronics, networks, and grids. The initiative also supports laboratory and curriculum development to integrate research and education.	$2,582,868
Sustainable Energy Pathways: This initiative, implemented by the Directorate for Engineering, is part of a broader agency initiative that supports innovative, interdisciplinary, basic research in science, engineering, and education by teams of researchers for developing systems approaches to sustainable energy pathways based on a comprehensive understanding of the scientific, technical, environmental, economic, and societal issues.	Information not available[b]
Renewable Energy Storage: This initiative, implemented by the Emerging Frontiers in Research and Innovation Office, supported basic research projects that invested in, and could potentially transform, the field of renewable energy storage.	6,000,000
Total	**$8,582,868**

Source: GAO analysis of agency-provided data.

[a]NSF officials reported that information on obligations for battery-related research awards was not available at the time of our review because the initiative's review/award process was still under way. They told us that funding for the Energy for Sustainability Program was about $13 million for fiscal year 2012. The program covers a variety of topics ranging from solar, wind, biofuels, and battery research. Officials told us that, based on the quality of the received proposals, and other factors, they anticipate there will be some battery-related research awards made during fiscal year 2012.

[b]NSF officials from this initiative reported that information on obligations in the area of batteries and energy storage technologies will not be known until awards are made.

Table 12: Environmental Protection Agency Battery and Energy Storage Initiatives and Their Total Funding Obligations for Fiscal Years 2009 through 2012

Name of initiative and description	Total (actual and estimated) obligations fiscal years 2009 through 2012
High-Pressure Accumulator Energy Storage for Hydraulic Hybrid Vehicles: This initiative, implemented by the Office of Transportation and Air Quality's Clean Automotive Technology Program, aims to develop a safe, low-cost, light weight, high-efficiency energy storage system for use in hydraulic hybrid passenger vehicles and heavy commercial trucks.	$3,258,029

Source: GAO analysis of agency-provided data.

Table 13: National Institute of Standards and Technology Battery and Energy Storage Initiatives and Their Total Funding Obligations for Fiscal Years 2009 through 2012

Name of initiative and description	Total (actual and estimated) obligations fiscal years 2009 through 2012
Development of Measurement Methods and Devices to Characterize Electrochemical Energy Storage and Conversion Devices at the Nanoscale: This single project, within the Center for Nanoscale Science and Technology, supports basic research to develop new tools and measurement systems to characterize the chemical and physical transformations that occur at the nanoscale in electrochemical energy storage devices.	$1,375,000[a]

Source: GAO analysis of agency-provided data.

[a]All or part of the agency-provided obligations are estimated.

Appendix IV: Selected Questionnaire Responses for Federal Battery and Energy Storage Technology Initiatives

We identified 39 initiatives across six agencies that supported batteries and other energy storage technologies through basic research, applied research, demonstrations, commercialization, and deployment activities. We developed a questionnaire about the initiatives and submitted it to the agencies. Tables 14, 15, 16, 17, 18, 19, and 20 provide selected questionnaire responses for the initiatives in each agency. We are reporting responses that elaborate on our report findings.

Table 14: Department of Energy Selected Questionnaire Responses for Battery and Energy Storage Initiatives

Key

1. Advanced lead-acid	12. Auxiliary power for vehicles	19. Basic research
2. Basic energy storage research	13. Ground-based vehicle propulsion	20. Applied research
3. Capacitors	14. Other vehicle propulsion	21. Demonstrations
4. Compressed air energy storage	15. Personal electronics	22. Commercialization
5. Flywheels	16. Stationary power storage	23. Deployment
6. Lithium-ion batteries	17. Weapon systems	24. Other
7. Lithium-metal batteries	18. Other	
8. Metal-air batteries		
9. Redox flow batteries		
10. Sodium batteries		
11. Other		

Initiative	Technologies											Uses							Technology advancement activities					
	1	2	3	4	5	6	7	8	9	10	11	12	13	14	15	16	17	18	19	20	21	22	23	24
OVT Vehicle Technologies Energy Storage Research and Development	✓		✓			✓	✓	✓				✓	✓							✓				
OE Energy Storage Program	✓		✓	✓	✓	✓	✓	✓	✓	✓						✓			✓	✓				
BES Core Research - Materials Science and Engineering		✓																	✓					✓
BES Energy Frontier Research Centers		✓																	✓					✓
BES Batteries and Energy Storage Energy Innovation Hub	✓	✓	✓			✓	✓	✓	✓	✓	✓			✓		✓			✓					✓

Key

1.	Advanced lead-acid	12.	Auxiliary power for vehicles	19.	Basic research
2.	Basic energy storage research	13.	Ground-based vehicle propulsion	20.	Applied research
3.	Capacitors	14.	Other vehicle propulsion	21.	Demonstrations
4.	Compressed air energy storage	15.	Personal electronics	22.	Commercialization
5.	Flywheels	16.	Stationary power storage	23.	Deployment
6.	Lithium-ion batteries	17.	Weapon systems	24.	Other
7.	Lithium-metal batteries	18.	Other		
8.	Metal-air batteries				
9.	Redox flow batteries				
10.	Sodium batteries				
11.	Other				

Initiative	Technologies											Uses							Technology advancement activities					
	1	2	3	4	5	6	7	8	9	10	11	12	13	14	15	16	17	18	19	20	21	22	23	24
ARPA-E Batteries for Electrical Energy Storage in Transportation			✓			✓		✓	✓		✓		✓					✓		✓		✓		
ARPA-E Grid-Scale Rampable Intermittent Dispatchable Storage	✓			✓	✓			✓	✓		✓					✓				✓		✓		
ARPA-E Initial Funding Opportunity Announcement – Energy Storage Technologies			✓			✓		✓			✓		✓			✓				✓		✓		
ARPA-E High Energy Advanced Thermal Storage										✓			✓			✓				✓		✓		
LPO Advanced Technology Vehicles Manufacturing Loan Program	✓					✓	✓	✓	✓	✓	✓		✓									✓	✓	
LPO Title XVII Loan Guarantee Program	✓		✓	✓	✓	✓	✓	✓	✓	✓	✓		✓			✓						✓		
Total	6	3	5	3	3	7	5	8	6	5	7	2	6	0	0	6	0	2	3	6	1	6	1	3

Source: GAO analysis of survey results.

Table 15: Department of Defense Selected Questionnaire Responses for Battery and Energy Storage Initiatives

Key

1.	Advanced lead-acid	12.	Auxiliary power for vehicles
2.	Basic energy storage research	13.	Ground-based vehicle propulsion
3.	Capacitors	14.	Other vehicle propulsion
4.	Compressed air energy storage	15.	Personal electronics
5.	Flywheels	16.	Stationary power storage
6.	Lithium-ion batteries	17.	Weapon systems
7.	Lithium-metal batteries	18.	Other
8.	Metal-air batteries	19.	Basic research
9.	Redox flow batteries	20.	Applied research
10.	Sodium batteries	21.	Demonstrations
11.	Other	22.	Commercialization
		23.	Deployment
		24.	Other

Initiative	Technologies											Uses							Technology advancement activities					
	1	2	3	4	5	6	7	8	9	10	11	12	13	14	15	16	17	18	19	20	21	22	23	24
Army Energy Storage for the Soldier		✓	✓			✓	✓	✓							✓		✓		✓	✓	✓		✓	
Army Energy Storage for Basing		✓				✓										✓			✓	✓				
Army Energy Storage for Air and Ground Vehicles	✓	✓	✓			✓	✓	✓				✓	✓						✓	✓				
Navy and Marine Corps Energy Storage Science and Technology		✓	✓			✓	✓	✓		✓		✓	✓	✓	✓	✓	✓		✓	✓	✓			
Navy and Marine Corps Energy Storage System Integration	✓		✓			✓	✓	✓		✓	✓	✓	✓	✓		✓	✓	✓		✓	✓		✓	
Air Force Basic Research Activities in Battery and Storage Technologies		✓																	✓					
Air Force Wide Temperature Capacitor Research and Development		✓	✓									✓		✓		✓				✓	✓			

Key

1. Advanced lead-acid
2. Basic energy storage research
3. Capacitors
4. Compressed air energy storage
5. Flywheels
6. Lithium-ion batteries
7. Lithium-metal batteries
8. Metal-air batteries
9. Redox flow batteries
10. Sodium batteries
11. Other

12. Auxiliary power for vehicles
13. Ground-based vehicle propulsion
14. Other vehicle propulsion
15. Personal electronics
16. Stationary power storage
17. Weapon systems
18. Other

19. Basic research
20. Applied research
21. Demonstrations
22. Commercialization
23. Deployment
24. Other

Initiative	Technologies											Uses							Technology advancement activities					
	1	2	3	4	5	6	7	8	9	10	11	12	13	14	15	16	17	18	19	20	21	22	23	24
Air Force Batteries for Aircraft and Directed Energy Weapons			✓			✓	✓	✓				✓		✓			✓			✓	✓	✓		
Air Force Zinc-Bromide Flow Battery	✓										✓		✓			✓					✓		✓	
Air Force Special Purpose Power						✓		✓						✓	✓				✓	✓		✓		
Air Force Advanced Materials for Energy Storage Applications		✓	✓			✓	✓	✓	✓		✓	✓		✓			✓		✓	✓				
Air Force Batteries for Space-Based Vehicles		✓				✓						✓		✓					✓					✓
DARPA Revolutionary Portable Energy Storage for the Warfighter		✓	✓			✓	✓	✓	✓				✓	✓	✓			✓	✓	✓	✓	✓	✓	✓
ESTCP Installation Energy Test Bed	✓					✓			✓	✓						✓				✓				
Total	**4**	**8**	**9**	**0**	**0**	**11**	**7**	**8**	**2**	**3**	**4**	**7**	**5**	**8**	**4**	**5**	**6**	**2**	**6**	**11**	**10**	**2**	**5**	**1**

Source: GAO analysis of survey results.

Table 16: National Aeronautics and Space Administration Selected Questionnaire Responses for Battery and Energy Storage Initiatives

Key

1. Advanced lead-acid
2. Basic energy storage research
3. Capacitors
4. Compressed air energy storage
5. Flywheels
6. Lithium-ion batteries
7. Lithium-metal batteries
8. Metal-air batteries
9. Redox flow batteries
10. Sodium batteries
11. Other

12. Auxiliary power for vehicles
13. Ground-based vehicle propulsion
14. Other vehicle propulsion
15. Personal electronics
16. Stationary power storage
17. Weapon systems
18. Other

19. Basic research
20. Applied research
21. Demonstrations
22. Commercialization
23. Deployment
24. Other

Initiative	Technologies											Uses							Technology advancement activities					
---	1	2	3	4	5	6	7	8	9	10	11	12	13	14	15	16	17	18	19	20	21	22	23	24
Prognostics Algorithm Development			✓			✓						✓		✓					✓	✓	✓			
Night Rover Challenge																				✓	✓			
Silicon Nano-Wire Anode						✓								✓				✓		✓	✓			
Space Power Systems Project						✓						✓	✓					✓		✓	✓			
Flywheel Energy Storage and Momentum Control					✓											✓		✓		✓	✓	✓		
Aerospace Lithium-Ion Cell Qualification Program						✓								✓						✓	✓		✓	
Lithium-Ion COTS Battery Surveillance						✓	✓	✓				✓	✓		✓					✓		✓	✓	✓
NASA Space Act Agreement with Underwriters Laboratories						✓	✓					✓	✓	✓	✓					✓	✓	✓	✓	✓
Total	0	0	1	0	1	6	2	1	0	0	0	4	3	4	2	1	0	3	1	8	7	3	3	2

Source: GAO analysis of survey results.

Table 17: National Science Foundation Selected Questionnaire Responses for Battery and Energy Storage Initiatives

Key

1. Advanced lead-acid	12. Auxiliary power for vehicles	19. Basic research
2. Basic energy storage research	13. Ground-based vehicle propulsion	20. Applied research
3. Capacitors	14. Other vehicle propulsion	21. Demonstrations
4. Compressed air energy storage	15. Personal electronics	22. Commercialization
5. Flywheels	16. Stationary power storage	23. Deployment
6. Lithium-ion batteries	17. Weapon systems	24. Other
7. Lithium-metal batteries	18. Other	
8. Metal-air batteries		
9. Redox flow batteries		
10. Sodium batteries		
11. Other		

Initiative	Technologies											Uses							Technology advancement activities					
	1	2	3	4	5	6	7	8	9	10	11	12	13	14	15	16	17	18	19	20	21	22	23	24
Energy for Sustainability		✓				✓	✓	✓					✓						✓					
Energy, Power, and Adaptive Systems	✓	✓	✓			✓	✓	✓	✓	✓	✓								✓					
Sustainable Energy Pathways											✓								✓					
Renewable Energy Storage		✓	✓	✓		✓					✓					✓			✓	✓				
Total	1	3	2	1	0	3	2	2	1	1	3	0	1	0	0	1	0	0	4	1	0	0	0	0

Source: GAO analysis of survey results.

Table 18: Environmental Protection Agency Selected Questionnaire Responses for Battery and Energy Storage Initiatives

Key

1.	Advanced lead-acid	12.	Auxiliary power for vehicles	19.	Basic research
2.	Basic energy storage research	13.	Ground-based vehicle propulsion	20.	Applied research
3.	Capacitors	14.	Other vehicle propulsion	21.	Demonstrations
4.	Compressed air energy storage	15.	Personal electronics	22.	Commercialization
5.	Flywheels	16.	Stationary power storage	23.	Deployment
6.	Lithium-ion batteries	17.	Weapon systems	24.	Other
7.	Lithium-metal batteries	18.	Other		
8.	Metal-air batteries				
9.	Redox flow batteries				
10.	Sodium batteries				
11.	Other				

Initiative	Technologies											Uses							Technology advancement activities					
	1	2	3	4	5	6	7	8	9	10	11	12	13	14	15	16	17	18	19	20	21	22	23	24
High-Pressure Accumulator Energy Storage for Hydraulic Hybrid Vehicles											✓	✓	✓			✓		✓		✓	✓	✓	✓	
Total	0	0	0	0	0	0	0	0	0	0	1	1	1	0	0	1	0	1	0	1	1	1	1	0

Source: GAO analysis of survey results.

GAO-12-842 Batteries and Energy Storage

Table 19: National Institute of Standards and Technology Selected Questionnaire Responses for Battery and Energy Storage Initiatives

Key

1. Advanced lead-acid	12. Auxiliary power for vehicles
2. Basic energy storage research	13. Ground-based vehicle propulsion
3. Capacitors	14. Other vehicle propulsion
4. Compressed air energy storage	15. Personal electronics
5. Flywheels	16. Stationary power storage
6. Lithium-ion batteries	17. Weapon systems
7. Lithium-metal batteries	18. Other
8. Metal-air batteries	
9. Redox flow batteries	
10. Sodium batteries	
11. Other	

19. Basic research	
20. Applied research	
21. Demonstrations	
22. Commercialization	
23. Deployment	
24. Other	

Initiative	Technologies											Uses							Technology advancement activities					
	1	2	3	4	5	6	7	8	9	10	11	12	13	14	15	16	17	18	19	20	21	22	23	24
Development of Measurement Methods and Devices to Characterize Electrochemical Energy Storage and Conversion Devices at the Nanoscale						✓													✓					
Total	0	0	0	0	0	1	0	0	0	0	0	0	0	0	0	0	0	0	1	0	0	0	0	0

Source: GAO analysis of survey results.

Table 20: Total for All Agencies—Selected Questionnaire Responses for Federal Battery and Energy Storage Initiatives

Key

1. Advanced lead-acid	12. Auxiliary power for vehicles	19. Basic research
2. Basic energy storage research	13. Ground-based vehicle propulsion	20. Applied research
3. Capacitors	14. Other vehicle propulsion	21. Demonstrations
4. Compressed air energy storage	15. Personal electronics	22. Commercialization
5. Flywheels	16. Stationary power storage	23. Deployment
6. Lithium-ion batteries	17. Weapon systems	24. Other
7. Lithium-metal batteries	18. Other	
8. Metal-air batteries		
9. Redox flow batteries		
10. Sodium batteries		
11. Other		

Agency	Technologies											Uses							Technology advancement activities					
	1	2	3	4	5	6	7	8	9	10	11	12	13	14	15	16	17	18	19	20	21	22	23	24
DOE	6	3	5	3	3	7	5	8	6	5	7	2	6	0	0	6	0	2	3	6	1	6	1	3
DOD	4	8	9	0	0	11	7	8	2	3	4	7	5	8	4	5	6	2	6	11	10	2	5	1
NASA	0	0	1	0	1	6	2	1	0	0	0	4	3	4	2	1	0	3	1	8	7	3	3	2
NSF	1	3	2	1	0	3	2	2	1	1	3	0	1	0	0	1	0	0	4	1	0	0	0	0
EPA	0	0	0	0	0	0	0	0	0	0	1	1	1	0	0	1	0	1	0	1	1	1	1	0
NIST	0	0	0	0	0	1	0	0	0	0	0	0	0	0	0	0	0	0	1	0	0	0	0	0
Total	**11**	**14**	**17**	**4**	**4**	**28**	**16**	**19**	**9**	**9**	**15**	**14**	**16**	**12**	**6**	**14**	**6**	**8**	**15**	**27**	**19**	**12**	**10**	**6**

Source: GAO analysis of survey results.

Appendix V: GAO's Questionnaire for Federal Agencies with Battery and Energy Storage Initiatives

 **United States Government Accountability Office**

QUESTIONS ABOUT FEDERAL BATTERY AND ENERGY STORAGE TECHNOLOGY-RELATED INITIATIVES: [PRE-POPULATED INFORMATION]

Introduction

The United States Government Accountability Office (GAO), an independent, legislative branch agency, is examining federal government initiatives to support basic and applied research, demonstration, and commercialization of battery and energy storage technologies (GAO job code 361332). Battery and energy storage technology-related initiatives are those that could or do support such technologies, either exclusively or as part of a broader initiative. GAO is undertaking this work at the request of the Chairman of the House Committee on Science, Space, and Technology. This work has three objectives: (1) to identify the key characteristics of battery and energy storage technology-related initiatives supported by federal agencies; (2) to determine the extent of any potential fragmentation, overlap, or duplication of federal battery and energy storage-related initiatives; and (3) to examine the extent to which federal agencies coordinate to achieve common goals for their battery and energy storage-related initiatives.

The purpose of this questionnaire is to collect information on key characteristics of your battery and energy storage technology-related initiative, funding, and coordination activities during fiscal years 2009 through 2012. Our review focuses on a variety of rechargeable (i.e. secondary) batteries and other energy storage technologies that include, but are not limited, to capacitors, flywheels, and compressed air energy storage (CAES). The term "technologies" encompasses full systems, as well as subsystems, components, and power management approaches for these energy storage technologies; for example battery components, battery cells and packs, and battery power management systems. Our review excludes nonrechargeable (i.e. primary) batteries, fuel cells, and radioisotope thermoelectric generators.

After we receive your response, we will follow up with pertinent agency official(s), if needed, to discuss and clarify any outstanding questions about the initiative as they relate to the objectives stated above.

Please complete all four sections and return it to one of the individuals named below by **March 30, 2012.** When returning the questionnaire, please attach any relevant supporting documentation to your e-mail.

If you have any questions or comments about this questionnaire, please contact Perry Lusk at (202) 512-3731 or Luskp@gao.gov. Thank you very much for your assistance.

Page 1 of 15

Appendix V: GAO's Questionnaire for Federal
Agencies with Battery and Energy Storage
Initiatives

QUESTIONS ABOUT FEDERAL BATTERY AND ENERGY STORAGE TECHNOLOGY-RELATED INITIATIVES- [PRE-
POPULATED INFORMATION]

Instructions

This questionnaire can be completed using MS-Word and returned via e-mail to
Luskp@gao.gov. Please complete this questionnaire and return it by **March 30, 2012**.

1) Please provide the information listed at the end of Section I for the battery and energy
   storage technology-related initiative.

2) Use your mouse to navigate by clicking on the box or check box ☐ you wish to answer.

3) To select a check box, click on the center of the box, and an 'X' will appear.

4) To deselect a check box response, click on the center of the box, and the 'X' will disappear.

5) To answer a question that requires a comment, click on the answer box and begin typing.
   The box will expand to accommodate your answer.

Page 2 of 15

SECTION I: GENERAL INFORMATION FOR INITIATIVE

We identified your agency and office as having a battery and energy storage technology-related initiative. We are interested in your initiative's activities from fiscal years 2009 through fiscal year 2012.

We understand that different agencies may use different terms to describe the various topics that we discuss in this questionnaire. For the purposes of this questionnaire, we have developed the following definitions:

- *Battery and energy storage technologies include, but are not limited to, rechargeable (i.e. secondary) batteries, capacitors, and flywheels. The term "technologies" encompasses full systems, as well as subsystems, components, and power management approaches for these energy storage technologies; for example, battery components, battery cells and packs, and battery power management systems. Our review excludes nonrechargeable (i.e. primary) batteries, fuel cells, and radioisotope thermoelectric generators.*
- *Initiative is a group of agency activities serving a similar purpose or function, such as a program or mission area.*
- *Battery and energy storage technology-related initiatives are those initiatives that could or do support battery and energy storage technologies, either exclusively or as part of a broader initiative.*
- *Individual projects exist as part of an overall initiative—for example, specific grant awards, or agreements, in-house research activities, or contracts that are supported by your initiative.*
- *Technology advancement activities may involve one or more of the following activities:*
  - *Basic research includes efforts to explore and define scientific or engineering concepts or is conducted to investigate the nature of a subject without targeting any specific technology.*
  - *Applied research includes efforts to develop new scientific or engineering knowledge to create new and improved technologies.*
  - *Demonstration activities include efforts to operate new or improved technologies to collect information on their performance and assess their readiness for widespread use.*
  - *Commercialization includes efforts to bridge the gap between research and demonstration activities, and venture capital funding and marketing activities, through transitioning technologies to commercial applications.*
  - *Deployment includes efforts to facilitate or achieve widespread use of technologies either in the commercial market or for federal agencies' use.*

We have excluded from our review any battery and energy storage technology-related initiatives supporting solely deployment. In addition, we are not reviewing activities involving agency-owned assets such as fleets or facilities.

Page 3 of 15

QUESTIONS ABOUT FEDERAL BATTERY AND ENERGY STORAGE TECHNOLOGY-RELATED INITIATIVES- [PRE-POPULATED INFORMATION]

Please provide the information listed below for the battery and energy storage technology-related initiative.

Initiative Title:

Implementing Agency:

Implementing Office:

Description of Initiative:

SECTION II: KEY CHARACTERISTICS OF INITIATIVE

For the purposes of this questionnaire, we consider an initiative to be a group of agency activities serving a similar purpose or function, such as a program or mission area. An initiative is considered "active" if it was planned or funded or implemented or authorized in the fiscal year described.

1) We understand that your initiative may support battery and energy storage technology advancement activities as well as other activities. Was this initiative active—for either battery and energy storage OR any other activities—at your agency at any time during fiscal years 2009 through 2012?

	Yes	No
a) FY 2009	☐	☐
b) FY 2010....................	☐	☐
c) FY 2011.....................	☐	☐
d) FY 2012.....................	☐	☐

If you checked No to any of the fiscal years above, please explain in the space below why the initiative was not active in that year. *For example, if the initiative was defunded.*

Page 4 of 15

QUESTIONS ABOUT FEDERAL BATTERY AND ENERGY STORAGE TECHNOLOGY-RELATED INITIATIVES- [PRE-POPULATED INFORMATION]

2) We understand that each agency defines technology advancement activities in a unique manner; however, in order to compare activities across agencies, we have developed definitions that can be applied broadly and have listed them below. Please determine which definition(s) best fit your individual initiative and answer for **all** of the initiative's activities, if applicable—not only for those that are battery and energy storage-related.

Does your initiative support any of the following technology advancement activities? Please check Yes or No for each activity.	Yes ▼	No ▼
a) Basic research .... *Basic research includes efforts to explore and define scientific or engineering concepts or is conducted to investigate the nature of a subject without targeting any specific technology.*	☐	☐
b) Applied research .... *Applied research includes efforts to develop new scientific or engineering knowledge to create new and improved technologies.*	☐	☐
c) Demonstration activities .... *Demonstration activities include efforts to operate new or improved technologies to collect information on their performance and assess their readiness for widespread use.*	☐	☐
d) Commercialization... *Commercialization includes efforts to bridge the gap between research and demonstration activities, and venture capital funding and marketing activities, through transitioning technologies to commercial applications.*	☐	☐
e) Deployment... *Deployment includes efforts to facilitate or achieve widespread use of technologies either in the commercial market or for federal agencies' use.*	☐	☐
f) Other *[Please specify below]*	☐	☐

Page 5 of 15

QUESTIONS ABOUT FEDERAL BATTERY AND ENERGY STORAGE TECHNOLOGY-RELATED INITIATIVES- [PRE-POPULATED INFORMATION]

3) **Does your initiative support any of the following battery and energy storage technologies?**

Battery and energy storage technologies encompass full systems, as well as subsystems, components, and power management approaches for these energy storage technologies; for example, development of rechargeable (i.e. secondary) battery components, battery cells and packs, and battery power management systems.

Please check Yes or No for each activity.

a) Rechargeable (i.e. secondary) batteries.... ☐ Yes ──────► **Does your initiative support any of the following types of rechargeable batteries?**
☐ No

	Yes	No
a) Advanced lead-acid batteries...........	☐	☐
b) Lithium-ion batteries....	☐	☐
c) Lithium-metal batteries...	☐	☐
d) Metal-air batteries........	☐	☐
e) Redox flow batteries....	☐	☐
f) Sodium batteries...........	☐	☐
g) Other batteries.............	☐	☐

b) Basic energy storage research..................... ☐ Yes ☐ No

c) Capacitors.................. ☐ Yes ☐ No

d) Compressed air energy storage (CAES)......... ☐ Yes ☐ No

e) Flywheels................... ☐ Yes ☐ No

f) Pumped hydro storage...................... ☐ Yes ☐ No

If you answered Other batteries, please describe:

g) Other

If you answered Other, please describe in the space below.

Appendix V: GAO's Questionnaire for Federal
Agencies with Battery and Energy Storage
Initiatives

QUESTIONS ABOUT FEDERAL BATTERY AND ENERGY STORAGE TECHNOLOGY-RELATED INITIATIVES- [PRE-POPULATED INFORMATION]

4) **We understand that some initiatives support basic research, which may not have a particular battery and energy storage technology *use* in mind. Does your initiative support specific technology uses?** *Please see the table below for examples of uses.*

Yes ....☐

No ....☐

If you answered No, please explain the types of research your initiative supports in the space below:

▨

If you answered Yes, which of the following technology uses, if any, does your initiative support?

Please check Yes or No for each technology use:

		If you checked Yes, please describe examples of the technology uses
a) Personal-use electronics power..................... *Internal power for small mobile electronic devices; for example, laptops, cell phones, and handheld radios.*	☐ Yes→ ☐ No	▨
b) Large-scale stationary power storage........... *Energy storage for stationary electricity generation and distribution systems that provide power to a large area for a large number of users; for example, the U.S. power grid or large military installations.*	☐ Yes→ ☐ No	▨
c) Small-scale stationary power storage......... *Energy storage for stationary electricity generation and distribution systems that provide power to a small area for a limited number of users; for example, forward operating bases or microgrids.*	☐ Yes→ ☐ No	▨
d) Ground-based vehicle propulsion............. *Power for propulsion of vehicles that travel along the ground, whether on earth or on other planets; for example electric and hybrid cars, tanks, exploration rovers.*	☐ Yes→ ☐ No	▨
e) Propulsion for other vehicles..................... *Power for propulsion of vehicles that travel in the air, space, underwater, or on the surface of the water; for example aircraft, submarines, and boats.*	☐ Yes→ ☐ No	▨
f) Auxiliary power for vehicles..................... *Power, for uses other than propulsion, for vehicles; for example, navigation, communication, and other equipment.*	☐ Yes→ ☐ No	▨
g) Weapon systems....................................... *Power for weapons and components necessary for their operation, such as targeting and guidance devices.*	☐ Yes→ ☐ No	▨
h) Other...	☐ Yes→ ☐ No	▨

Page 7 of 15

Appendix V: GAO's Questionnaire for Federal
Agencies with Battery and Energy Storage
Initiatives

QUESTIONS ABOUT FEDERAL BATTERY AND ENERGY STORAGE TECHNOLOGY-RELATED INITIATIVES- [PRE-
POPULATED INFORMATION]

5) **Does your initiative directly perform technology advancement activities, such as
research, for batteries and energy storage in-house?**

Yes..............☐
No..............☐

6) **Does your initiative provide assistance to recipients to perform technology advancement
activities for batteries and energy storage?**

Yes................☐
No................☐

If you answered Yes, what type(s) of assistance does your initiative provide? Please
check Yes or No for each activity.

	Yes	No
a) Cooperative agreements	☐	☐
b) Contracts...............................	☐	☐
c) Grants..................................	☐	☐
d) Direct loans............................	☐	☐
e) Interagency agreements.................	☐	☐
f) Loan guarantees........................	☐	☐
g) Tax credits.............................	☐	☐
h) Other..................................	☐	☐

If you answered Other, please describe the other type(s) of assistance that your initiative
provides in the space below.

Page 8 of 15

QUESTIONS ABOUT FEDERAL BATTERY AND ENERGY STORAGE TECHNOLOGY-RELATED INITIATIVES- [PRE-POPULATED INFORMATION]

7) **What type(s) of recipients are eligible to receive assistance from your initiative?**

Please check Yes or No for each activity:

	Yes	No
a) Department of Defense labs..............	☐	☐
b) Department of Energy national labs......	☐	☐
c) Other federal government labs...........	☐	☐
d) Industry....................................	☐	☐
e) Universities..............................	☐	☐
f) Other..	☐	☐

If you answered Other, please describe the other type(s) of recipients that your initiative funds in the space below:

8) **As part of our reporting to Congress, we would like to provide a few examples of projects that have been funded. In 800 or fewer characters, please briefly describe in the space below two or three examples of specific battery and energy storage technology-related projects funded or supported by this initiative:**

    a. Project #1

    b. Project #2

    c. Project #3

Page 9 of 15

Appendix V: GAO's Questionnaire for Federal
Agencies with Battery and Energy Storage
Initiatives

QUESTIONS ABOUT FEDERAL BATTERY AND ENERGY STORAGE TECHNOLOGY-RELATED INITIATIVES- [PRE-POPULATED INFORMATION]

SECTION III: GOALS AND PERFORMANCE MEASURES FOR INITIATIVE

9) Please describe your initiative's mission and overarching goals:

10) Has your initiative established goals that are <u>directly related to battery and energy storage technology advancement</u>?

    Yes........................☐
    No ....☐ (SKIP TO 12)

If you answered Yes, please describe the goals for battery and energy storage technology advancement in the space below:

11) Has your initiative created performance measures to track progress towards the <u>battery and energy storage technology advancement</u> goals?

    Yes ....☐
    No...........................☐

If you answered Yes, please describe the performance measures in the space below:

If you answered Yes, please describe the extent to which your initiative has met its goals during the period fiscal years 2009 through 2012:

QUESTIONS ABOUT FEDERAL BATTERY AND ENERGY STORAGE TECHNOLOGY-RELATED INITIATIVES- [PRE-POPULATED INFORMATION]

12) **Does your initiative have specific** *cost targets* **for the battery and energy storage technologies supported?**

Yes .... ☐
No.......................... ☐

If you answered Yes, please describe your initiative's cost targets in the space below:

If you answered Yes, to what extent has your initiative met the cost targets since fiscal year 2009? Please explain in the space below:

13) **Does your initiative have specific** *technical performance targets* **for the battery and energy storage technologies supported?**

Yes .... ☐
No.......................... ☐

If you answered Yes, please describe your initiative's technical performance targets in the space below:

If you answered Yes, to what extent has your initiative met the targets since fiscal year 2009? Please explain in the space below:

Page 11 of 15

Appendix V: GAO's Questionnaire for Federal
Agencies with Battery and Energy Storage
Initiatives

QUESTIONS ABOUT FEDERAL BATTERY AND ENERGY STORAGE TECHNOLOGY-RELATED INITIATIVES- [PRE-
POPULATED INFORMATION]

SECTION IV: FUNDING FOR INITIATIVE

We are interested in how much money goes toward the battery and energy storage technology-
related activities that your initiative supports. We would like to be able to report obligations data
across all federal battery and energy storage technology-related initiatives for fiscal years 2009
through 2012. If necessary, please consult with staff in your agency's budget office to answer
these funding-related questions. *Note: We may request copies of supporting documentation for
the numbers you provide below.*

For questions in this section, please use the following definition:
*An **obligation** is a definite commitment that creates a legal liability of the government for the
payment of goods and services ordered or received, or a legal duty on the part of the United
States that could mature into a legal liability. Payment may be made immediately or in the
future. An agency incurs an obligation, for example, when it places an order, signs a contract,
awards a grant, purchases a service, or takes other actions that require the government to make
payments to the public or from one government account to another.*

14) **For fiscal years 2009 through 2012, please provide actual or estimated obligations data
for your <u>initiative's battery and energy storage technology activities only</u>.**

	Total obligations for initiative's **battery and energy storage technology activities**	Check if the total is an estimate	Check if information is not available
a) Fiscal year 2009...............................	S____	☐	☐
b) Fiscal year 2010...............................	S____	☐	☐
c) Fiscal year 2011...............................	S____	☐	☐
d) Fiscal year 2012 *(estimated)*....................	S____		☐

If the total is an estimate, please describe how the estimate was determined and describe
any limitations associated with this estimate:

FY2009: █████

FY2010: █████

FY2011: █████

FY2012: █████

Page 12 of 15

QUESTIONS ABOUT FEDERAL BATTERY AND ENERGY STORAGE TECHNOLOGY-RELATED INITIATIVES- [PRE-POPULATED INFORMATION]

e) If actual or estimated obligations are not available, please explain why in the space below. (If you did provide obligations data, please proceed to Q14.)

FY2009:

FY2010:

FY2011:

FY2012:

Questions 15 through 16 ask about specific projects funded or supported by your initiative in fiscal years 2009 through 2012.

15) **How many total projects were active in this initiative in fiscal years 2009 through 2012?**
For the purposes of this questionnaire, we consider individual projects to exist as part of an overall initiative—for example, specific grant awards, or agreements, in-house research activities, or contracts that are supported by your initiative. We consider an initiative to be "active" if the initiative was planned or funded or implemented or authorized in the fiscal year described.

FY2009:

FY2010:

FY2011:

FY2012:

16) **We understand that some initiatives may support projects that are not battery and energy storage technology-related. Were all projects listed in question 15 battery and energy storage-related?**

Yes .... ☐

No .... ☐

If you answered No, how many of the total projects listed above were battery and energy storage technology -related and were active in this initiative in fiscal years 2009 through 2012?

FY2009:

FY2010:

FY2011:

FY2012:

Page 13 of 15

Appendix V: GAO's Questionnaire for Federal
Agencies with Battery and Energy Storage
Initiatives

QUESTIONS ABOUT FEDERAL BATTERY AND ENERGY STORAGE TECHNOLOGY-RELATED INITIATIVES- [PRE-
POPULATED INFORMATION]

SECTION V: COORDINATION

Congress and GAO are interested in understanding how agencies coordinate internally and with
other federal agencies on battery and energy storage technology-related initiatives.

17) **Do you or staff from your initiative *formally* coordinate with other battery and energy
storage technology-related initiatives within your agency or with other federal
agencies?** An example of formal coordination might be participation in a working group or
having an interagency agreement.

Yes .... ☐
No........................... ☐

If you answered Yes, please provide up to five examples of how you coordinate either within
your agency or with other agencies on your initiative. Please also provide electronic copies of
relevant documents or provide link(s) to relevant website(s) in the space below.

Name/title of coordination activities	Description of coordination activities	Does this coordination activity involve coordinating with other agencies?	
		Yes ▼	No ▼
1.		☐	☐
2.		☐	☐
3.		☐	☐
4.		☐	☐
5.		☐	☐

If you have additional information you would like to provide about your coordination activities,
please describe in the space below:

Page 14 of 15

18) **When reviewing applications for assistance, do you assess other sources of federal funding the potential recipient (i.e., applicant) has obtained for the same project?**

- Yes .... ☐
- No .... ☐
- Not Applicable .... ☐

If you answered Yes, please describe what information you review and how you assess the information, in the space below.

SECTION VI: OTHER INFORMATION

19) **Are there any additional data, further sources of information, nuances, or comments not covered previously that would help us further understand and report on how this initiative is being implemented?** *If you do not have additional information to provide, it is appropriate to leave this question blank.* If you do have additional information, please describe it in the space below and/or provide link(s) to the relevant website(s) in the space below or please provide electronic copies of the relevant documents (e.g., strategic planning documents, performance reports, requests for proposals, regulatory provisions, etc.) when returning this questionnaire.

Alternatively, you can suggest that further discussion about this initiative be conducted through follow-up conversations with GAO.

20) **Please provide a point of contact in your initiative for any follow-up questions we may have about the responses to this questionnaire:**

Agency/Organization:

Contact Name:

Email:

Phone:

Please remember to attach any relevant supporting documentation when returning this questionnaire to LuskP@gao.gov.

Thank you for your time!

Page 15 of 15

Appendix VI: GAO Contact and Staff Acknowledgments

GAO Contact	Frank Rusco (202) 512-3841 or ruscof@gao.gov
Staff Acknowledgments	In addition to the contact named above, Tim Minelli (Assistant Director), Hilary Benedict, Frederick K. Childers, Tanya Doriss, R. Scott Fletcher, Brian M. Friedman, Cindy Gilbert, Perry Lusk, Cynthia Norris, Jerome Sandau, Kathryn Smith, Maria Stattel, Barbara Timmerman, and Eugene Wisnoski made key contributions to this report.